Graduate Texts in Physics

Graduate Texts in Physics publishes core learning/teaching material for graduate- and advanced-level undergraduate courses on topics of current and emerging fields within physics, both pure and applied. These textbooks serve students at the MS- or PhD-level and their instructors as comprehensive sources of principles, definitions, derivations, experiments and applications (as relevant) for their mastery and teaching, respectively. International in scope and relevance, the textbooks correspond to course syllabi sufficiently to serve as required reading. Their didactic style, comprehensiveness and coverage of fundamental material also make them suitable as introductions or references for scientists entering, or requiring timely knowledge of, a research field.

Martin Stutzmann · Christoph Csoklich

The Physics of Renewable Energy

Springer

Martin Stutzmann
Walter Schottky Institute
Technical University Munich
Garching, Bayern, Germany

Christoph Csoklich
Electrochemistry Laboratory
Paul Scherrer Institut
Villigen, Switzerland

ISSN 1868-4513 ISSN 1868-4521 (electronic)
Graduate Texts in Physics
ISBN 978-3-031-17723-1 ISBN 978-3-031-17724-8 (eBook)
https://doi.org/10.1007/978-3-031-17724-8

Cover image: The Sun's south pole as seen by the ESA/NASA Solar Orbiter spacecraft. These images were recorded by the Extreme Ultraviolet Imager (EUI) at a wavelength of 17 nanometers. Permission for reproduction is gratefully acknowledged.

This Springer imprint is published by the registered company Springer Nature Switzerland AG
The registered company address is: Gewerbestrasse 11, 6330 Cham, Switzerland

We dedicate this book to a sustainable future of our planet Earth.
May it remain an inhabitable home for our children, grandchildren, and many future generations to come!
Indeed there is no planet B!

Preface

This book is based on the script of a lecture about renewable energy held regularly at the Physics Department of Technische Universität München since 2007. Over the years, the contents of this lecture were updated and extended to the present version. Although originally intended for master students in engineering physics, the lecture was also frequently followed by bachelor students and students of other faculties interested in this topic. That is why the requirements for pre-existing knowledge in physics vary from chapter to chapter. Some topics can be appreciated with a general scientific background, while others require the knowledge of more advanced physical concepts. In this way, the book hopefully appeals to a larger community of potential readers.

The main aim of the book is to make interested scientists aware of the scientific backgrounds, the potentials, but also the fundamental limitations of different renewable energy technologies in terms of achievable energy densities, maximum physical efficiencies, and future scientific and technological challenges. Whenever possible, we have tried to keep units, notations, and nomenclature consistent across the different forms of renewable energies, and to refrain from unphysical prejudices and biases.

We gratefully acknowledge the constructive feedback and questions of many hundred students over the years and the support of colleagues in this field by making their results available for this book. And we thank PA for just being it!

Munich, Germany
Villigen, Switzerland
July 2022

Martin Stutzmann
Christoph Csoklich

Contents

Acronyms

ADP	Adenosine diphosphate
AM	Air mass
APS	Artificial photosynthesis
ATP	Adenosine triphosphate
AU	Astronomical unit
BHJ	Bulk heterojunction
BSF	Back surface field
CAES	Compressed air energy storage
CB	Conduction band
CCS	Carbon capture and storage
Chl	Chlorophyll
CIS	Copper-indium-diselenide
CSP	Concentrated solar power
Cys	Cysteine
Cyt bf	Cytochrome bf
DOS	Density of states
DSSC	Dye-sensitized solar cell
EF	Fermi level
ETP	Electron transfer phosphorylization
Fd	Ferredoxin
FF	Fill factor
HAWT	Horizontal axis wind turbine
HDR	Hot dry rock
HOMO	Highest occupied molecular orbital
IHP	Inner Helmholtz plane
ITCZ	Intertropical conversion zone
ITER	International thermonuclear experimental reactor
ITO	Indium-tin oxide
kgCE	Kilogram of coal equivalent
kgOE	Kilogram of oil equivalent
LHC	Light-harvesting complex
LO	Longitudinal optical
LUMO	Lowest unoccupied molecular orbital
MIS	Metal-Insulator-Semiconductor

MPP	Maximum power point
NADP	Nicotineamid adenine dinucleotid phosphate
NHE	Normal hydrogen electrode
OEC	Oxygen-evolving center
OHP	Outer Helmholtz plane
OPV	Organic photovoltaics
PCy	Plastocyanin complex
PECVD	Plasma-enhanced chemical vapor deposition
PEFC	Polymer electrolyte membrane fuel cell
Ph a	Phaeophytin a
PQ	Plastoquinone
PS I	Photosystem I
PS II	Photosystem II
Pyr	Pyrrole ring
RC	Reaction center
SC	Solar constant
SHE	Standard hydrogen electrode
SOFC	Solid oxide fuel cell
TCO	Transparent conducting oxide
TEG	Thermoelectric generator
VB	Valence band

Energy—A Brief Introduction 1

Abstract

This chapter deals with a theoretical introduction to the physical meaning of energy and work. The principles of energy conversion from one form to another are explained, as well as the impossibility of energy "destruction". After some thermodynamic considerations about energy, the Noether theorem is shortly explained. The chapter finishes with common units of energy and some numbers, to get a first impression of the magnitudes mentioned in this book.

The physical description of our world starts from the fundamental coordinates in space, $\mathbf{r}$, and time, t. For a moving object, the space vector is a function of time, $\mathbf{r}(t)$, defining the trajectory of the object with the velocity $\mathbf{v}(t)$ as the time derivative of $\mathbf{r}(t)$ (Fig. 1.1). Closely linked to the velocity is the momentum $\mathbf{p} = m \cdot \mathbf{v}$ of an object with mass m. The space and momentum vectors form a pair of so called *conjugated coordinates*, which in quantum mechanics gives rise to the first Heisenberg uncertainty relationship between space and momentum. Also the time t has such a conjugated coordinate, namely the energy E, giving rise to the time-energy uncertainty relationship in quantum mechanics. As we will briefly discuss further below, the well-known conservation laws for momentum and energy also are the result of the invariance of our world with respect to translations of the corresponding conjugated coordinates space and time.

Of course, the concept of energy was developed long before modern theoretical physics and quantum mechanics. The word *energy* comes from the Greek expression *en-érgeia* ($\epsilon\nu$-$\epsilon\rho\gamma\epsilon\iota\alpha$) which can be translated as *acting force*. A more modern version of this definition would be (Fig. 1.1):

> Energy is the variable of state of a physical system describing the ability of the system to perform work on other physical systems via an interaction due to forces between the systems.

M. Stutzmann and C. Csoklich, *The Physics of Renewable Energy*,
Graduate Texts in Physics,
https://doi.org/10.1007/978-3-031-17724-8_1

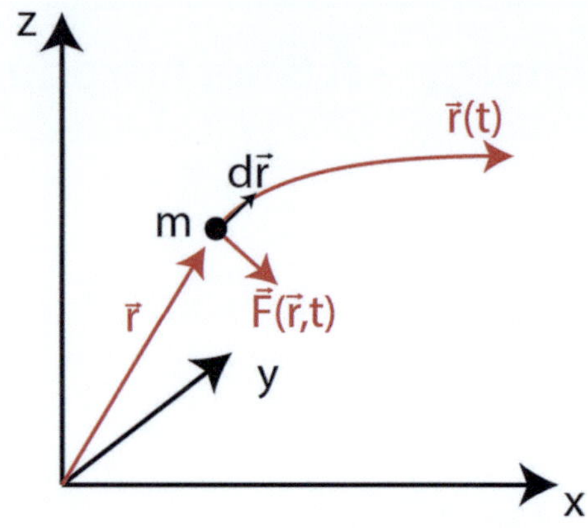

Fig. 1.1 A point mass m at position $\mathbf{r}$ moves along the trajectory $\mathbf{r}(t)$ and changes its original momentum due the acting force $\mathbf{F}(\mathbf{r}, t)$

1.1 Energy and Work

The change in energy E of a system is given by the work W performed by an external force $\mathbf{F}(r, t)$ acting on this system (cf. Fig. 1.1):

$$dW = \mathbf{F}(\mathbf{r}, t) \cdot d\mathbf{r} \quad \Rightarrow \quad W_{1,2} = \int_{\mathbf{r}_1}^{\mathbf{r}_2} \mathbf{F}(\mathbf{r}, t) d\mathbf{r} \tag{1.1.1}$$

Depending on the sign of dW, two qualitatively different cases can occur:

- $dW > 0$: if the external force has a component parallel to $d\mathbf{r}$, dW is positive and the energy of the system increases by transferring energy from a second, external system from which the force originates.
- $dW < 0$: in this case, the energy of the system decreases by performing work on a second system.

1.2 Potential and Kinetic Energy

Since according to the discussion above the nature of the interaction force determines how the energy of a system will be changed, we will have a closer look at the kind of forces encountered in nature. Here, one distinguishes between conservative and non-conservative or dissipative forces. A force is called conservative, if the work performed by it along any closed trajectory is 0:

$$W = \oint_{\mathbf{r}_1}^{\mathbf{r}_1} \mathbf{F}(\mathbf{r}, t) \cdot d\mathbf{r} = 0 \tag{1.2.1}$$

In this special case, the work $W_{1,2}$ performed by the conservative force on a trajectory between the point $\mathbf{r}_1$ and $\mathbf{r}_2$ will be independent of the specific trajectory chosen. Then we can define a **potential energy** E_{pot} such that

$$\Delta E_{pot} = E_{pot}(\mathbf{r}_2) - E_{pot}(\mathbf{r}_1) = W_{1,2} \tag{1.2.2}$$

$$\mathbf{F} = -\nabla \cdot E_{pot} \tag{1.2.3}$$

with the gradient vector operator ∇:

$$\nabla = \text{grad} = \left(\frac{\partial}{\partial x}, \frac{\partial}{\partial y}, \frac{\partial}{\partial z}\right)$$

Note that only the change ΔE_{pot} can be related to the measurable physical quantity $W_{1,2}$, so that the origin of the potential energy scale can be chosen arbitrarily (or conveniently). Moreover, the potential energy provides a convenient way to reconstruct the interaction force as the negative gradient of E_{pot}. Unfortunately, most forces encountered in reality are non-conservative, e.g. due to friction or turbulences.

A second interesting case is a force which accelerates a mass according to Newtons second law $\mathbf{F} = m\mathbf{a}$. Then, a simple calculation leads to the definition of a **kinetic energy** E_{kin}:

$$\Delta E_{kin} = \int_{\mathbf{r}_1}^{\mathbf{r}_2} \mathbf{F}(\mathbf{r}, t) \cdot \mathrm{d}\mathbf{r} = \int_{\mathbf{r}_1}^{\mathbf{r}_2} m \left(\frac{\mathrm{d}\mathbf{v}}{\mathrm{d}t}\right) \cdot \mathrm{d}\mathbf{r} = \tag{1.2.4}$$

$$\int_{\mathbf{r}_1}^{\mathbf{r}_2} m\mathrm{d}\mathbf{v} \cdot \frac{\mathrm{d}\mathbf{r}}{\mathrm{d}t} = \int_{\mathbf{r}_1}^{\mathbf{r}_2} m\mathbf{v} \cdot \mathrm{d}\mathbf{v} = \frac{1}{2}mv^2(\mathbf{r}_2) - \frac{1}{2}mv^2(\mathbf{r}_1) \tag{1.2.5}$$

If we now choose the origin of the kinetic energy as 0 in $\mathbf{r}_1$, we can define the kinetic energy as:

$$\mathbf{v}_1 = 0 : \Rightarrow \; E_{kin} = \frac{1}{2}mv^2 \tag{1.2.6}$$

As an important result, the Noether theorem discussed below states that in systems with only dissipation-less, conservative forces the total mechanical energy E is conserved:

$$E = E_{kin} + E_{pot} = \text{const.} \tag{1.2.7}$$

1.3 Noether Theorem and Energy Conservation

The conservation of energy is one of many conservation laws in physics. According to the general theorem formulated by the mathematician Emmy Noether in 1918, conservation laws are a consequence of symmetries (invariant operations) in a physical system [1]. The symmetries and corresponding conserved physical quantities are summarized in Table 1.1.

Table 1.1 Symmetries and corresponding conserved physical quantities

Invariance	Conserved quantity
Translation in 3D space	Momentum $m\mathbf{v}$
Translation in a periodic lattice	Quasi-momentum $\hbar\mathbf{k}$ [a]
Translation in time	Energy E
Rotation in 3D	Angular momentum $m\mathbf{r} \times \mathbf{v}$
Mirror operation	Parity
Phase shift of a quantum-mechanical wave function	Electric charge q

[a] wavevector **k**

Here, we just give a brief sketch of this theorem as applied to an invariance of a system under translation in time. The argument is based on the Lagrange formalism in classical mechanics, where a mechanical system is described by the Lagrange operator $\mathcal{L}(q_i, \dot{q}_i)$ with generalised coordinates q_i and $\dot{q}_i = \mathrm{d}q_i/\mathrm{d}t$, i.e. the total derivative of q_i with respect to time. The Lagrange operator is obtained from the kinetic energy T and the potential energy V as $\mathcal{L} = T - V$. For the simplest case of a one-dimensional motion of a mass m this gives $\mathcal{L}(x, \dot{x}) = \frac{1}{2} \cdot m\dot{x}^2 - V(x)$. Using the variational principle of Hamilton, this leads to the Euler-Lagrange equation as the generalized equation of motion:

$$\frac{\mathrm{d}}{\mathrm{d}t}\left(\frac{\partial \mathcal{L}}{\partial \dot{q}_i}\right) = \frac{\partial \mathcal{L}}{\partial q_i} \tag{1.3.1}$$

For our simple one-dimensional example, this just gives the Newton equation of motion, $\mathbf{F} = m\mathbf{a}$. Now consider the following coordinate transformation:

$$q_i(t) \quad \rightarrow \quad q_i(s,t) \quad \text{with} \quad q_i(0,t) = q_i(t) \tag{1.3.2}$$

Taking the partial derivative of the Euler-Lagrange equation (1.3.1) with respect to s and rearranging, this gives:

$$\underbrace{\frac{\partial}{\partial s}\mathcal{L}\left(q_i(s,t), \dot{q}_i(s,t)\right)}_{\mathrm{I}} = \underbrace{\frac{\mathrm{d}}{\mathrm{d}t}\left(\frac{\partial \mathcal{L}\left(q_i(s,t), \dot{q}_i(s,t)\right)}{\partial \dot{q}_k(s,t)} \frac{\partial q_k(s,t)}{\partial s}\right)}_{\mathrm{II}} \tag{1.3.3}$$

If the transformation is a symmetry operation which leaves the system invariant, then the left side (part I) must be zero, meaning that expression II on the right side is constant in time, i.e. is preserved. For a translation in time, the coordinate transformation to be considered is $q_i(s,t) = q_i(s+t)$. From the Noether theorem then follows:

$$\frac{d}{dt}\left(\frac{\partial \mathcal{L}\left(q_i(s,t),\dot{q}_i(s,t)\right)}{\partial \dot{q}_k(t)}\dot{q}_k(t) - \mathcal{L}\left(q_i(t),\dot{q}_i(t)\right)\right) = 0 \qquad (1.3.4)$$

Here, the expression within the brackets is the total energy of the system, which accordingly is conserved. This again can be seen using the simple example of a free mass m without a potential, $V(x) = 0$:

$$\mathcal{L} = \frac{1}{2}m\dot{x}^2; \qquad \frac{\partial \mathcal{L}}{\partial \dot{x}}\dot{x} - \mathcal{L} = m\dot{x}^2 - \frac{1}{2}m\dot{x}^2 = \frac{1}{2}m\dot{x}^2 = E_{kin} \qquad (1.3.5)$$

1.4 Inner Energy

In general, processes on earth are non-conservative and irreversible due to friction, turbulences, mixing, etc. Then, on a macroscopic level, energy conservation seems to be violated, because the mechanical energy $E_{kin} + E_{pot}$ is no longer constant. Instead, it is partially transformed into inner energy E_I (or heat Q) stored in the interacting atoms of a macroscopic system. This inner energy is the sum over the (kinetic and potential) energy of every single atom with respect to the center of gravity $\mathbf{R}_S$ of the system:

$$E_I = \sum_i E_i(\mathbf{r} - \mathbf{R}_S) \qquad (1.4.1)$$

The conversion of macroscopic potential or kinetic energy to inner energy by dissipation can occur to a 100% (e.g. the kinetic energy of a moving car is completely transformed to heat by the brakes), while the conversion of inner energy or heat back to macroscopic energy is limited by the laws of thermodynamics, in particular by the Carnot efficiency of heat engines. Thus, the second law of thermodynamics states that it is impossible to realise a cyclic heat engine which transforms heat into work with an efficiency of 100%. This fact motivates the distinction of two components of energy with different properties for practical use:

Exergy is the part of the total energy of a system which can be converted into useable macroscopic energy or work.

Anergy is the part of the total energy which cannot be converted into useable work.

For reversible processes, the relative amounts of exergy and anergy are conserved. Irreversible processes involving friction, turbulences, etc. cause an irreversible conversion of exergy to anergy, meaning a loss in useable energy.

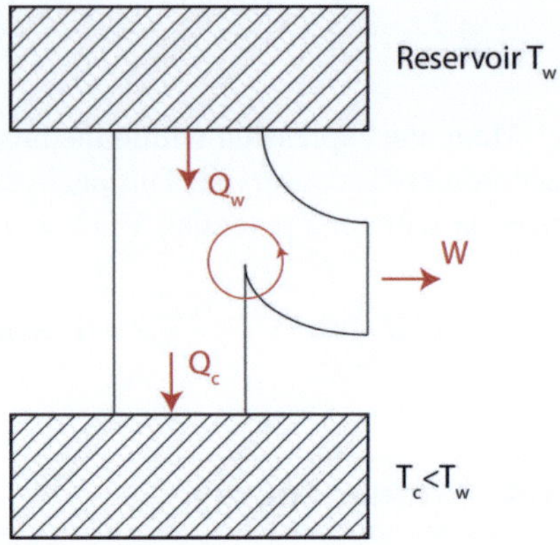

Fig. 1.2 Schematic energy flow in a cyclic heat engine. A cyclical process takes a heat Q_W from a hot reservoir at an absolute temperature T_W. Part of this heat is transformed into usable work W, whereas the rest flows as waste heat Q_C into a cold reservoir at temperature $T_C < T_W$

Consider a simple case: When a stone falls down onto the Earth from a height h_0, its original macroscopic potential energy $E_{pot} = mgh_0$ is transformed continuously into macroscopic kinetic energy E_{kin}. If we assume negligible friction with the surrounding air during the fall, there will be no increase of the initial temperature T_0 of the stone and, thus, no energy dissipation into the inner energy of the stone. Thus, during the fall, all atoms of the stone will have on average the same velocity $\mathbf{v}$ as the center of gravity of the stone. When the stone hits the ground in an ideal inelastic way, E_{kin} becomes abruptly 0 (due to $\mathbf{v} = 0$) and is completely transformed into inner energy E_I. This corresponds to an increase in temperature: $T_1 > T_0$ so that $E = mgh_0 = E_I$. This increased temperature T_1 of the stone could be used by a heat engine to lift the stone back up. However, due to the limited efficiency of the engine, the final reachable height h_1 will be smaller than the original height h_0 from which the stone has fallen.

The efficiency of the heat engine shown in Fig. 1.2 is limited by the Carnot efficiency:

$$\eta = \frac{|W|}{|Q_W|} = \frac{|Q_W| - |Q_C|}{|Q_W|} \leq \frac{T_W - T_C}{T_W} \tag{1.4.2}$$

For irreversible processes the anergy is lost as an the increase of entropy ($dS = dQ/T$)

$$E_{an} = T_C \Delta S = T_C \frac{Q_C}{T_C} = Q_C \tag{1.4.3}$$

1.5 Quantifying Energy

The derived SI unit for energy is the Joule:

$$J = kg\,m^2\,s^{-2} \tag{1.5.1}$$

However, depending on the scientific or practical context, a wide variety of other energy units is still in use. This is also due to the vast range of energy magnitudes that need to be described. Very common in physics is the unit electron volt (eV), to quantify e.g. energy scales for atoms or fundamental particles, whereas the energy produced by the Sun per second (around 45 magnitudes higher) can be better quantified in yottajoule, (YJ). The calorie (cal), now mostly used in the description of the energy content of food, was also popular in chemistry and is sometimes still in use. In engineering some other energy units, such as e.g. the kilowatt hour (kW h), are common (1 kW h corresponds to about 3.6×10^6 J). In the general public, quite often a confusion exists between energy and power: power is the change in energy per unit time, leading to its unit, the Watt (W).

$$W = J\,s^{-1} \tag{1.5.2}$$

The common confusion of kilowatt (a power unit) and kilowatt hour (an energy unit) should be carefully avoided. The Tables 1.2 and 1.3 show the relations of the most common physical/chemical units. Other units common in energy technology are the British Thermal Unit 1 BTU = 1055 J, the (kilogram of) Oil Equivalent 1 kgOE = 4.2×10^7 J = 11.63 kW h, (kilogram of) Coal Equivalent 1 kgCE = 2.93×10^7 J = 8.141 kW h, or the horse power 1 PS = 735.5 W as a publicly used unit of power. On the global scale, the terawatt year (TWa) is another useful energy unit.

Table 1.4 shows a brief overview of how much energy is consumed currently on different scales by human beings compared to the energy typically available from

Table 1.2 Important physical and chemical units of energy

	J	eV	cal
J	1	6.24×10^{18}	0.239
eV	1.602×10^{-19}	1	3.83×10^{-12}
cal	4.1855	2.61×10^{19}	1

Table 1.3 Important technical units of energy

	kW h	TWa	kgOE	kgCE
kW h	1	1.14×10^{-13}	8.60×10^{-2}	0.123
TWa	8.76×10^{12}	1	7.53×10^{11}	1.08×10^{12}
kgOE	11.63	1.33×10^{-12}	1	1.43
kgCE	8.141	9.28×10^{-13}	0.7	1

Table 1.4 Comparison of the magnitudes of energy consumption and renewable energy supply on Earth

	Power/Energy
Basic metabolic rate of a human being	80 W = 700 kWh/a = 86 kgCE/a
Basic metabolic rate of the world population	7×10^{11} kgCE/a
Average **energy consumption** in Germany/capita	5 kW = 45 MWh/a = 5500 kgCE/a
Total **energy consumption** per year	
GER (82 M inhabitants)	5×10^{11} kgCE
US (330 M)	3×10^{12} kgCE
Chad (10 M)	9×10^{8} kgCE = 90 kgCE/capita
Total **solar irradiation**	1.7×10^{17} W = 1.8×10^{17} kgCE/a
Including **direct radiation energy**	3×10^{16} kgCE/a (18%)
Wind, water, waves	5×10^{16} kgCE/a (30%)
Biomass	2×10^{14} kgCE/a (0.1%)
Earth heat	3.5×10^{13} kgCE/a (0.02 %)
Tidal wave energy	3.5×10^{12} kgCE/a (0.002 %)
Total **renewable energy production** in GER	6×10^{10} kgCE/a (≈12%)

different renewable energy sources. The minimum power to keep us alive is our basic metabolic rate of about 80 W [2]. Over the time of a year, this sums up to an energy demand of 700 kWh/a or 86 kgCE/a. Thus, we would need about our body weight in coal per year to simply exist. For the current Earth's population of about 8 billion people, the accumulated minimal energy demand would be 7×10^{11} kgCE/a. However, especially in rich countries much more energy is consumed to satisfy our current standards in mobility, health, culture or luxury. Thus, in Germany we presently use on the average a constant power of 5 kW per capita instead of the 80 W we really need [3]. In other countries like the US this total power consumed per person is even larger (currently around 10 kW). Only in very poor countries like the Chad the actual power demand is close to the basic metabolic rate.

On the other side, the Sun supplies us with an enormous power of solar irradiation of continuously more than 2×10^{17} W. The solar energy arriving on Earth within one hour would be enough to satisfy our current global energy demand for an entire year. Of course, not all of this incoming solar energy is readily available for our energy supply. Almost 50% of the incoming radiation is radiated back into space by reflection at the atmosphere or thermal radiation of the Earth. But the other 50% could be harvested by us using direct solar radiation via solar cells, wind engines, water power plants, biomass etc. Compared to that, other renewable energy sources like geothermal energy or tidal energy are almost negligible. But overall, there is basically no physical limitation to satisfy our energy demands from renewable energy sources if we really want to. This means that there is still a lot of room for improving the current contribution of about 12% of renewable energy to the total energy production in e.g. Germany, and the same holds for the entire Earth as a whole. To convey a feeling of which physical restrictions we will have to beat or live with in this challenge is what this little book is all about.

Further Reading

- Our world in data: research and data to make progress against the world's largest problems: ourworldindata.org. Accessed from 26 July 2022.
- Data on the German electricity marked: smard.de. Accessed from 26 July 2022.
- Worldwide total primary energy supply (TPES) by source: www.iea.org/data-and-statistics/. Accessed from 26 July 2022.
- BP Statistical Review of World Energy: https://www.bp.com/en/global/corporate/energy-economics/statistical-review-of-world-energy.html. Accessed from 30 July 2022.

References

1. Noether, E.: Invariante Variationsprobleme, Nachrichten von der Gesellschaft der Wissenschaften zu Göttingen. Mathematisch-Physikalische Klasse **1918**, 235–257 (1918)
2. Mifflin, M.D., et al.: A new predictive equation for resting energy expenditure in healthy individuals. Am. J. Clin. Nutr. **51**, 241–247 (1990)
3. BP: Statistical Review of World Energy 2021, vol. 70 (2021)

Forms of Energy and Their Density

2

Abstract

In this chapter, we recall briefly different forms of energy in view of their potential for energy storage or transmission. Although most physics students will be familiar with most of the formulae below for the different energies E, they will be much less familiar as far as the related achievable volumetric energy densities E/V are concerned. So it is worthwhile to have a closer look at this.

2.1 Mechanical Potential Energy

As already introduced in Sect. 1.2, potential energy can be defined in systems subjected to a conservative force field. The gravitational field is conservative to a very good approximation and, thus, a potential energy between different points in the gravitational force field can be defined. In the simplest case this gives:

$$E_{pot} = m \cdot g \cdot h \tag{2.1.1}$$

Here m denotes the mass of an object in the Earth's gravitational field, g the gravitational acceleration,[1] and h the height of the object above ground as the arbitrarily chosen zero point (h should be small enough so that g still can be considered constant).

[1] In the following examples we use the approximated value for the gravitational acceleration at the Earth's surface of $g = 9.81\,\mathrm{m\,s^{-2}}$.

M. Stutzmann and C. Csoklich, *The Physics of Renewable Energy*,
Graduate Texts in Physics,
https://doi.org/10.1007/978-3-031-17724-8_2

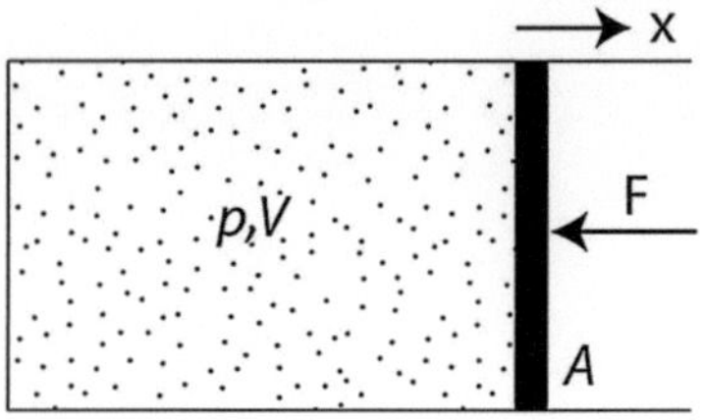

Fig. 2.1 Illustration of the volume work required to compress gas

As an example, 1 t of water is carried up a mountain to a height of 100 m:

$$E_{pot} = mgh \approx 1 \times 10^6\,\mathrm{J}$$

This is equivalent to 0.3 kWh/250 kcal/0.04 kgCE and corresponds to an energy density (assuming $\rho = 0.998\,\mathrm{g\,cm^{-3}}$ for the density of water) of only

$$E/V = 1 \times 10^6\,\mathrm{J\,m^{-3}} \quad (2.1.2)$$

As a consequence, the energy density in pumped hydro-storage is quite low. Therefore, reservoirs require huge volumes and big enough falling heights to achieve significant storage capabilities.

Another storage system using the potential energy difference between two states is the compressed air energy storage (CAES). The compression and expansion of air is used to take up excess energy e.g. from a wind park and return it to the electrical grid when demanded. For this, a turbine operates in connection with an underground cavern.

The compression of an ideal gas follows the ideal gas equation:

$$pV = NkT = \nu RT \quad (2.1.3)$$

with the mol number ν, the Boltzmann constant k, and the gas constant R. For compression we have to perform volume work on the system (cf. Fig. 2.1):

$$\mathrm{d}W = -F\mathrm{d}x = -pA\mathrm{d}x = -p\mathrm{d}V$$

If a gas is very well insulated from the environment, then the compression occurs adiabatically: the heat ΔQ exchanged between the gas and the surrounding is ≈ 0. This is a good approximation for a CAES system, since Earth is a bad thermal conductor.

For a more detailed calculation of the adiabatic compression, we start from the first law of thermodynamics:

$$dU = \delta Q + \delta W \tag{2.1.4}$$

Here, U is the inner energy of the gas, W is the mechanical work $W = \int_{V_1}^{V_2} -p dV$ upon volume change, Q is the heat, here defined by the heat capacity C: $\Delta Q = C \cdot \Delta T$. More exactly, the heat capacity depends on whether the inner energy U is changed under the condition of constant volume or constant pressure:

$$C_V = \left.\frac{dU}{dT}\right|_V = \frac{3}{2}kN = \frac{3}{2}\nu R \qquad \text{and} \tag{2.1.5}$$

$$C_P = \left.\frac{dU}{dT}\right|_P = \frac{5}{2}kN = \frac{5}{2}\nu R = C_V + \nu R \tag{2.1.6}$$

For an adiabatic compression of the gas we then have:

$$\begin{aligned} \delta Q = 0 = dU - \delta W = C_V dT + p dV \quad & \Big| \; p = \frac{\nu RT}{V} \\ C_V dT + \nu RT \frac{dV}{V} = 0 \quad & \Big| \; : C_V T \\ \frac{dT}{T} + \frac{\nu R}{C_V}\frac{dV}{V} = 0 \quad & \Big| \; \frac{\nu R}{C_V} = \frac{C_P - C_V}{C_V} = \gamma - 1 \end{aligned}$$

with γ, the *adiabatic coefficient*:

$$\gamma = \frac{C_P}{C_V} \tag{2.1.7}$$

Integration then gives:

$$\begin{aligned} \ln T + (\gamma - 1) ln V &= const \\ \ln(T V^{\gamma-1}) &= const \\ T V^{\gamma-1} &= const \quad \Big| \; T = \frac{pV}{\nu R} \end{aligned}$$

This leads to the equation of adiabaticity:

$$pV^{\gamma} = \text{const.} \tag{2.1.8}$$

Now consider the adiabatic compression of a gas from $\{T_0, p_0\}$ to the final condition $\{T_1, p_1\}$ with $r = p_1/p_0$ as the ratio of compression:

$$\left(\frac{T_1}{T_0}\right)^{\gamma} = \left(\frac{p_1}{p_0}\right)^{\gamma-1} \tag{2.1.9}$$

Thus, the amount of energy stored in the compressed gas is:

$$E = C_V(T_1 - T_0) = C_V T_0 \left(\frac{T_1}{T_0} - 1\right) = C_V T_0 (r^{1-\frac{1}{\gamma}} - 1) \tag{2.1.10}$$

The adiabatic coefficient of dry air is $\gamma = 1.4$. We start from a temperature of $T_0 = 0\,°\mathrm{C}$ and a pressure of $p_0 = 1\,\mathrm{bar}$. The heat capacity of air is $C_V = 21\,\mathrm{J\,K^{-1}\,mol}$, where 1 mol of air at the starting conditions has a volume of 22.4 L. From this, we can calculate now the energy density:

$$\frac{E}{V} = 0.07(r - r^{\frac{1}{\gamma}})\mathrm{kW\,h\,m^{-3}} \tag{2.1.11}$$

This expression depends only on the ratio of compression r. Assuming a moderate value of $r = 30$ for the compression ratio, we obtain an energy density of:

$$E/V = 1.3\,\mathrm{kW\,h\,m^{-3}} = 5 \times 10^6\,\mathrm{J\,m^{-3}} \tag{2.1.12}$$

To store 1 GW h we then need a volume of $10^6\,\mathrm{m}^3$, which corresponds to a relatively small cubic cavern of size $100\,\mathrm{m} \times 100\,\mathrm{m} \times 100\,\mathrm{m}$.

2.2 Kinetic Energy

As discussed in Sect. 1.2, kinetic energy is the energy stored in the linear motion of an object, described by the translational velocity v:

$$E_{kin} = \frac{1}{2}mv^2 \tag{2.2.1}$$

As an example, consider a car with a mass of $m = 1000\,\text{kg}$, driving with a speed of $v = 100\,\text{km}\,\text{h}^{-1} = 28\,\text{m}\,\text{s}^{-1}$. This gives a kinetic energy of:

$$E_{kin} = 4 \times 10^5\,\text{J} \approx 0.1\,\text{kW}\,\text{h} = 8.6\,\text{gOE}$$

So, the entire kinetic energy of a fast-moving car is only equivalent to the energy content a few mL of oil. Similar to our example for potential energy above, this again shows how little energy is contained in a mechanical form compared to a chemical form of energy.

In addition, also rotating extended objects have kinetic energy in the form of rotational energy. Using $v = \omega \cdot r$ for the velocity of a mass element rotating with angular velocity ω at a distance r around a fixed rotation axis, integration over the rotating object gives for the rotational energy:

$$E_{rot} = \frac{1}{2} I \omega^2 \tag{2.2.2}$$

Here, I is the moment of inertia of the rotating object obtained by integration over all mass elements.

An interesting example is the rotational energy of the Earth. Approximating the Earth by a homogeneous sphere, we have $I = 2/5mr^2$, with $m = 5.9 \times 10^{24}\,\text{kg}$, and $r = 6.4 \times 10^6\,\text{m}$. The angular velocity is given by $\omega = 2\pi/24\,\text{h} = 7.3 \times 10^{-5}\,\text{s}^{-1}$. Inserting this into the above equation, we find :

$$E_{rot} = 2.5 \times 10^{29}\,\text{J} = 8 \times 10^9\,\text{TWa} \tag{2.2.3}$$

This is a huge energy when compared to the total energy consumption of mankind of about 15 TWa. However, it is very difficult to use this energy source for our needs. This can only be accomplished by tidal energies as discussed in Chap. 4.

Rotational energy storage is practically used in flywheel farms e.g. for server facilities as a quickly accessible backup energy source. Similarly, the linear kinetic energy of a moving car could be transferred to or from a flywheel storage system during braking or acceleration. The requirements to store the kinetic energy of our car above in a flywheel can be easily calculated:

Assume a flywheel in form of a cylindrical torus with mass $m = 50\,\text{kg}$ and radius $r = 50\,\text{cm}$. To store the linear kinetic energy of the moving car discussed above, this flywheel would have to rotate at about $\omega = 270\,\text{s}^{-1} \approx 2500\,\text{rpm}$, a value which could cause serious problems for the driving behaviour of the car because of unwanted gyroscopic effects of the flywheel.

2.3 Wave Energy

In a mechanical oscillator like a pendulum, potential and kinetic energy are periodically converted into each other. This also is the case for waves in coupled mechanical oscillators. Such waves have the interesting property that they can transport energy from one place to another without a corresponding transport of mass. We will have a closer look at water and tidal waves in Chap. 4. Here we just briefly summarize some basic aspects of their energy content.

2.3.1 Mechanical Waves

Mechanical waves like water waves have an energy density per unit wave front given by (cf. Sect. 4.1):

$$E = \frac{1}{2}\rho_{lin}\omega^2 A^2 l \tag{2.3.1}$$

Here, ρ_{lin} is the linear mass density of the medium (in units [kg m^{-1}]), A is the wave amplitude and l the length of the wave front.

A water wave with amplitude $A = 1\,\text{m}$ and frequency $f = 0.5\,\text{Hz}$ has a linear power density (per m wave front) of:

$$\frac{dP}{dx} \approx 2000\,\text{W m}^{-1} \tag{2.3.2}$$

This may appear a lot for a human being enjoying waves at a beach, but for technical use this is again a very small density: to construct a wave power plant with a total power of 1 GW (i.e. the power of a typical nuclear plant) we would need 500 km of coastline.

2.3.2 Electromagnetic Waves

Electromagnetic waves, especially in the form of microwaves or light, are an interesting technical alternative to transport energy (and information) without a mechanical connection or an electric transmission line. After all, the Earth is powered by the Sun in this way, as we will discuss in detail in Sect. 3.1. From the Maxwell equations we find for the energy density of an infinite electromagnetic plane wave with the electric field $\boldsymbol{\mathcal{E}}$ and magnetic field $\boldsymbol{\mathcal{H}}$:

$$\frac{E}{V} = \frac{1}{2}\varepsilon\varepsilon_0\mathcal{E}^2(t) + \frac{1}{2}\mu\mu_0\mathcal{H}^2(t) = \frac{1}{2}(\boldsymbol{\mathcal{E}} \cdot \boldsymbol{\mathcal{D}} + \boldsymbol{\mathcal{H}} \cdot \boldsymbol{\mathcal{B}}) \tag{2.3.3}$$

To transmit a power of 1000 mW through vacuum or air with a microwave beam at a frequency of 2.45 GHz, the electrical field strength has to be approximately $1 \times 10^4\,\mathrm{V\,m^{-1}}$.

2.4 Electrostatic and Magnetostatic Energy

Also static electric and magnetic fields can store a considerable amount of energy. For electric fields $\boldsymbol{\mathcal{E}}$, this can be most easily seen by considering a plate capacitor with capacitance C which is charged up to a voltage U, corresponding to a total stored charge $q_{tot} = C \cdot U$. By integration from 0 to U, this gives for the stored energy:

$$E = \frac{1}{2}CU^2 = \frac{1}{2}q_{tot}U \tag{2.4.1}$$

Using the well known formula for the capacitance C of a parallel plate capacitor with plate area A and plate distance d and replacing the voltage U by the electric field $E = U/d$, we can easily calculate the energy density E/V:

$$C = \varepsilon\varepsilon_0\frac{A}{d} \tag{2.4.2}$$

$$\frac{E}{V} = \frac{1}{2}\varepsilon\varepsilon_0\boldsymbol{\mathcal{E}}^2 \tag{2.4.3}$$

Note that this agrees also with the energy density of the electrical part given above for electromagnetic waves. A very popular application of this form of energy storage are supercapacitors or ultracapacitors with a very high capacitance C. This can be achieved by using dielectric insulators with a high relative dielectric constant ε (high-k dielectrics), by minimizing the effective plate distance d (e.g. by making use of the Helmholtz layer at a metal/electrolyte interface, see below), and by maximizing the effective plate area A using dense arrays of conducting nanowires or nanoporous metallic foams instead of flat metal plates.

Consider a commercial supercapacitor with the dimensions 10 cm × 10 cm × 0.5 mm, a mass of 15 g and a capacitance of $C = 30$ F. This supercapacitor can be charged up very rapidly to a voltage of $U_{max} = 3$ V and then stores an energy of $E = 135$ J = 4 mgSCE. The corresponding energy density is then:

$$\frac{E}{V} = 3 \times 10^7 \, \mathrm{J\,m^{-3}} \tag{2.4.4}$$

Note, that this is 30 times higher than the energy density of a ton of water in a height of 100 m discussed in the beginning of this chapter.

In an analogous way we can calculate the energy stored in the static magnetic field $\mathcal{H}$ of a bobbin with N windings, length l and cross-section area A conducting an electric current I. The relevant physical quantities for this case are the inductance L of the bobbin and the magnetic flux ϕ generated by the flowing current I. Then the following relations hold:

$$E = \frac{1}{2}LI^2 = \frac{1}{2}\Phi I, \qquad L = \mu\mu_0 N^2 \frac{A}{l}, \qquad \Phi = N\mathcal{B}A \tag{2.4.5}$$

$$E = \frac{1}{2}\mathcal{B}^2 l A/(\mu\mu_0) = \frac{1}{2}\mu\mu_0 \mathcal{H}^2 V \tag{2.4.6}$$

$$\frac{E}{V} = \frac{1}{2}\mu\mu_0 \mathcal{H}^2 \tag{2.4.7}$$

Superconducting magnets have a $\mathcal{B}_{max}$ of ca. 20 T (1 T = 1 V s m^{-2}), limited by the critical magnetic field of the superconductor used. For $\mu = 1$ we find $\frac{E}{V} \leq 1.5 \times 10^8 \, \mathrm{J\,m^{-2}}$ (for comparison: the energy density of a Li-ion battery is about $10^9 \, \mathrm{J\,m^{-3}}$).

2.5 Latent Heat

Latent heat is a characteristic form of inner energy required for a phase change of condensed matter: melting of a solid or evaporation of a liquid as endothermic phase changes, and solidification of a liquid or condensation of a gas as the corresponding exothermic processes. A practical example is the use of water freezing to prevent the temperature in a room to fall below 0 °C. The water/ice phase transition has a latent heat of 335 J g^{-1}, corresponding to a volumetric energy density of $E/V =$

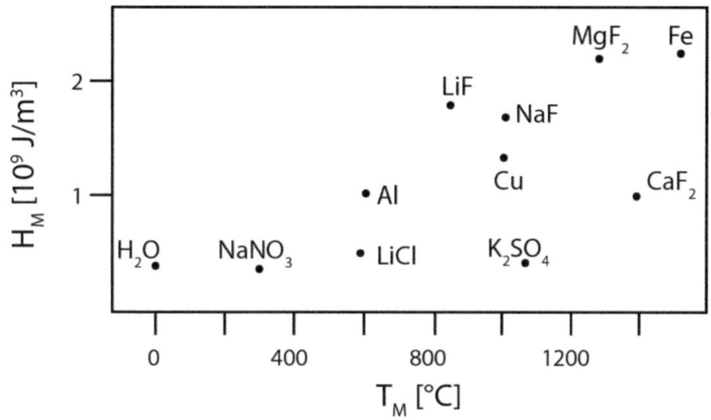

Fig. 2.2 Melting temperatures T_M and the corresponding latent heat densities H_M for different metals and compounds

$3.3 \times 10^8\,\mathrm{J\,m^{-3}}$. Thus, a 10 L bucket of water can provide a latent heat of about 1 kW h or 100 gCE upon freezing.

For more modern applications of latent heat energy storage e.g. in fuel cell housings or solar thermal power plants, other materials such as metals or special "phase change salts" (in particular metal fluorides) are used. The idea is to have the phase change occurring at a specific temperature level in order to thermally stabilize a system at this temperature, or to allow a high thermodynamic Carnot efficiency for the transformation of the stored latent heat into usable exergy. Figure 2.2 shows some examples of phase change materials for different applications.

2.6 Chemical and Electrochemical Energy

Chemical energy in the form of different fuels is the most common way in which we store and transport energy today. The energy in the fuel is liberated by exothermic chemical reactions, most commonly by oxidation with O_2 from the air (burning). The microscopic origin is the reduction of electronic potential energy of the reacting atoms or molecules. The simplest example is the formation of H_2 molecules from two hydrogen atoms shown in Fig. 2.3. We start from two single H atoms whose energy is defined as zero. For large distances r between these atoms, their interaction is negligible. When the atomic $1s$ wave functions start to overlap for small r, both electrons can lower their energy because they become less confined in space. When the equilibrium distance r_0 of the two protons in the H_2 molecule is reached, the stable bond is established with binding energy $E(r_0)$. The bonding orbital of the molecule is given by a symmetric combination of the two $1s$ orbitals with antisymmetric spins. It has a non-zero value of the squared electronic wave function between the two nuclei characteristic for a covalent bond. In contrast, the high energy antibonding level results from an antisymmetric combination of the two atomic wave functions with a characteristic zero-crossing between the nuclei and a symmetric spin alignment.

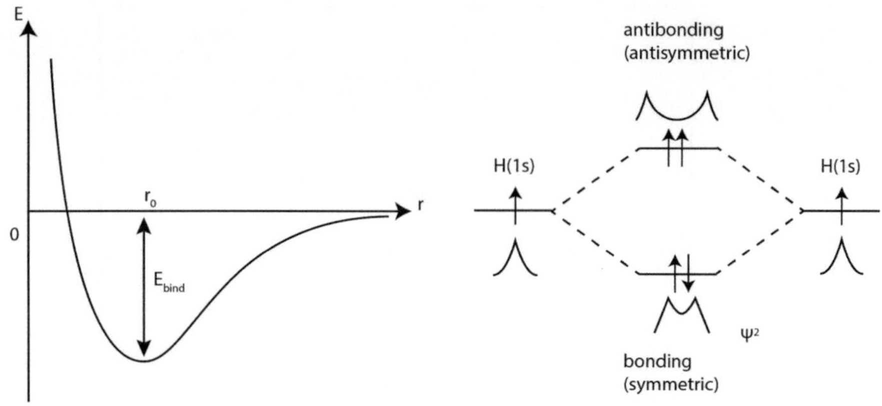

Fig. 2.3 Schematic view of the formation of a H_2 molecule from two hydrogen atoms. Left: Energy diagram of the total energy versus the internuclear distance r. The minimum corresponds to the binding energy stored in the bond and determines the equilibrium distance r_0 between the atomic nuclei after molecule formation. Right: Schematic sketch of the Ψ^2 wave functions and energy levels of the electrons in the atoms and in the molecule. Arrows indicate the electronic spins

In the molecular ground state, the two electrons from the two hydrogen atoms only occupy the bonding orbital. When one or both electrons are excited to the antibonding level e.g. by heat or optical excitation, the bond becomes unstable and the molecule can dissociate again.

The characteristic binding energy of a hydrogen molecule is 4.4 eV. Other important exothermic reactions are the burning of carbon or the oxidation of hydrogen to water:

$$H + H \longrightarrow H_2 + 4.4\,\text{eV}$$
$$C + O_2 \longrightarrow CO_2 + 4.2\,\text{eV}$$
$$H_2 + O \longrightarrow H_2O + 3\,\text{eV}$$

In a macroscopic chemical reaction, the binding energy is dissipated in the form of heat. This occurs through the interaction of a newly formed molecule with the environment via its translational, rotational and internal degrees of freedom. For the H_2 molecule discussed above, these degrees of freedom are the translations in the three orthogonal directions in space, rotation about the two orthogonal axes with finite momentum of inertia perpendicular to the bond axis, and the internal symmetric bond stretching vibration in the bond axis around the equilibrium distance r_0.

The unique advantage of chemical energy compared to all other forms of energy discussed so far is that the corresponding typical energy densities are very high:

For example, a hydrogen storage tank with compressed H_2 at 700 bar for the oxidation to water in a fuel cell has an energy density of:

$$E/V = 5 \times 10^9\,\mathrm{J\,m^{-3}}$$

Another order of magnitude higher is the energy density of gasoline for combustion in a car engine:

$$E/V = 5 \times 10^{10}\,\mathrm{J\,m^{-3}}$$

Of great practical importance is also the direct conversion of chemical energy into electrical energy, e.g. in batteries or fuel cells. Here, three different types of converters are distinguished:

primary electrochemical elements: non-chargeable batteries
secondary electrochemical elements: rechargeable batteries
tertiary electrochemical elements: with input of external masses, e.g. fuel cells

All these electrochemical elements consist of combinations of metal electrodes with electrolytes as ionic conductors. As shown in Fig. 2.4, at the metal/electrolyte interface positive or negative metal ions are dissolved to a certain extent in the electrolyte, while the electronic countercharge (electrons or holes (missing electrons)) remains in the metal. As a consequence, an electric field builds up between the electrode and the electrolyte, which increases until it prevents a further dissolution of metal ions and an equilibrium is obtained [1,2]. The interfacial electric field leads to

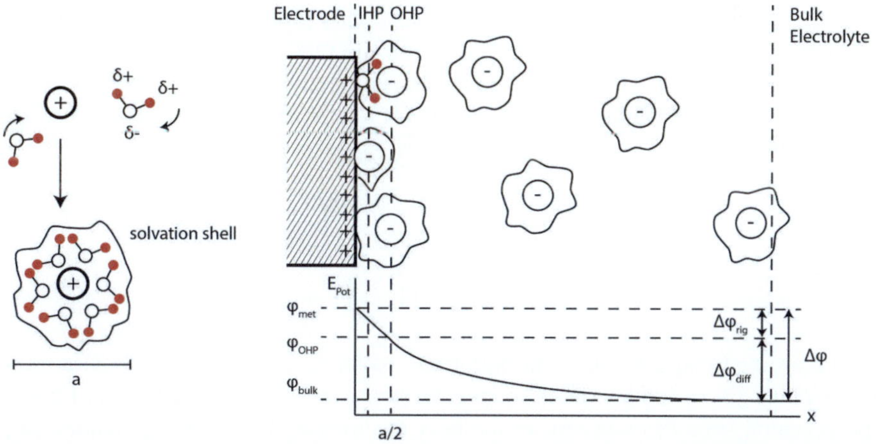

Fig. 2.4 The electrochemical double layer at an (metal) electrode/aqueous electrolyte interface. Ions in the electrolyte are surrounded by a solvation shell of electrolyte molecules, lowering their energy

Table 2.1 Galvanic series of some common elements

Element	Ox. State	U_0 [V]	Element	Ox. State	U_0 [V]
Au	1	+1,69	Fe	3	−0,04
Cl	1	+1,35	Pb	2	+0,13
Au	3	+1,40	Sn	2	−0,14
Pt	2	+1,18	Ni	2	−0,26
Ag	1	+0,80	Fe	2	−0,45
Graphite	2	+0,75	Zn	2	−0,76
Cu	1	+0,52	Al	3	−1,66
O_2	2	+0,39	Na	1	−2,71
Cu	2	+0,34	K	1	−2,93
Sn	4	+0,02	Li	1	−3,04
H_2	1	+−0,00			

a macroscopic potential difference between the metal electrode and the bulk of the electrolyte with a characteristic galvanic voltage U_0. This voltage is measured under standard conditions (room temperature, 1 M electrolyte) against a standard hydrogen electrode (SHE) as the zero point of the voltage scale. Table 2.1 lists the galvanic voltages for different metals and gases in different oxidation states.

Figure 2.4 provides a closer look at the structure and the potential profile in the so-called *electrochemical double layer* as a function of the distance x from the electrode surface. First, dissolved ions can adsorb directly at the electrode surface, defining the inner Helmholtz plane (IHP). In polar solvents like water, however, most dissolved ions are surrounded by a solvation shell of solvent molecules as indicated on the left side of the figure. If a is the diameter of the solvation shell, the so-called outer Helmholtz plane (OHP) due to adsorbed solvated ions occurs at $x = a/2$. In this region, the double layer can be described like a plate capacitor, so that the potential ϕ decreases linearly with distance x. Beyond the OHP, ions can diffuse, but still feel a finite voltage U, which influences their local concentration c and, therefore, the potential profile via the Nernst equation for ions with charge $\pm ze$ at an absolute temperature T:

$$c = c_0 \exp\left(-\frac{zeU}{kT}\right) \tag{2.6.1}$$

$$U = -\frac{kT}{ze} \ln\left(\frac{c}{c_0}\right) \tag{2.6.2}$$

According to Eq. 2.6.2, the concentration c of singly charged ions ($z = 1$) at room temperature ($T = 300$ K) varies by one decade when U varies by 59 mV. Finally, the potential becomes constant in the bulk electrolyte. The overall potential drop $\Delta\phi$ corresponds the measured or applied external voltage U between the electrode and the bulk electrolyte. The overall extent of the electrochemical double layer with varying potential depends on the ion concentration and the voltage and usually is as small as a few nanometers.

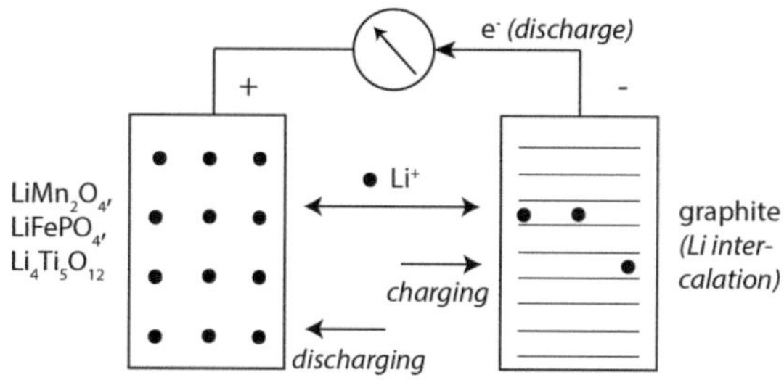

Fig. 2.5 Schematic view of a lithium ion battery and the directions of ionic and electronic transport during charging and discharging

By combination of two different electrodes and a suitable electrolyte, complete electrochemical elements can be constructed. As an example, we show below a schematic view of a rechargeable Lithium-polymer battery widely used today in portable devices. One electrode consists of a Li-rich metallic compound such as $LiMn_2O_4$, the other is made from graphite. Both electrodes are separated by a thin polymeric electrolyte through which Li^+ ions can move easily. When the battery is charged, an external voltage is applied which causes Li ions to flow from the metal to the graphite electrode, where they can intercalate easily between the stacked graphene layers. To maintain charge neutrality of each electrode, a corresponding amount of electrons is pumped from left to right through the external circuit. Upon discharge, this process is inverted as the ions move back to the metal electrode where they have a lower energy (Fig. 2.5).

A particular important property of batteries today is the amount of energy stored per unit volume or per unit mass. Both parameters should be as high as possible for practical use in portable electronics or e-mobility. Figure 2.6 shows the range of

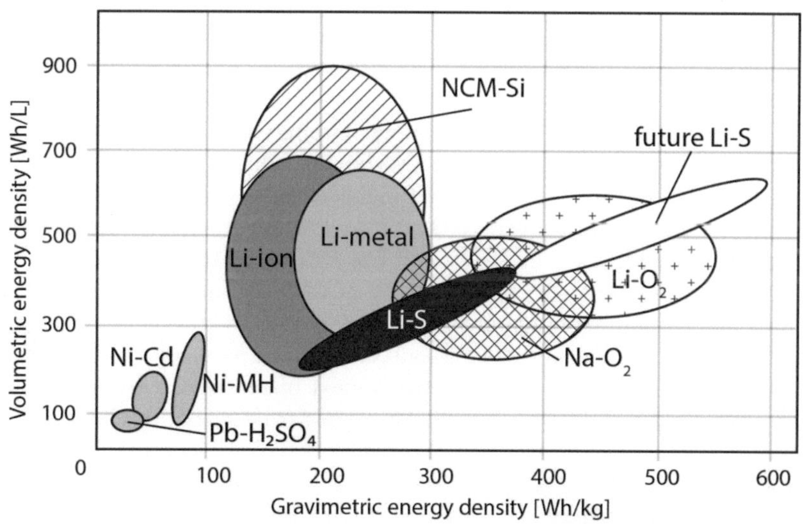

Fig. 2.6 Volumetric and gravimetric energy densities of past, present and future primary and secondary battery technologies (Ni–MH: nickel-metal hydride, NCM–Si: nickel-cobalt-manganese oxide-silicon)

Table 2.2 Properties of popular primary batteries

Type	Zinc manganese oxide	Zinc/air	Li-ion (non-rechargable)
Anode	$Zn + 4\,OH^-$	$2\,Zn + 8\,OH^-$	$xLiC_6$
	$\longrightarrow [Zn(OH)_4]^{2-} + 2\,e^-$	$\longrightarrow 2\,Zn(OH)_2\,{}_{-4} + 4\,e^-$	$\longrightarrow xLi^+ + xe^- + xC_6$
Cathode	$MnO_2 + H_2O + e^-$	$O_2 + 2\,H_2O + 4\,e_-$	$Li_{1-x}CoO_2 + xLi^+ + xe^-$
	$\longrightarrow MnO(OH) + OH^-$	$\longrightarrow 4\,OH^-$	$\longrightarrow LiCoO_2$
Energy density	$0.5\,MJ\,kg^{-1}$	$1.59\,MJ\,kg^{-1}$	$0.875\,MJ\,kg^{-1}$
Cell voltage application	1.5 V standard alkaline battery	1.4 V hearing devices	3.6 V mobile and automotive applications

Table 2.3 Comparison of two commercially available fuel cell systems [4]

	PEFC[a]	SOFC[b]
Operating temperature	<100 °C	900–1000 °C
Efficiency	60%	60%
Fuel	H_2, methanol, etc.	H_2, CO, CH_4, etc.
Fuel reforming	External	Internal
Anode material	Pt, Pt − X (X = Ru, Ni, Co)	Ni − YSZ-Cermet
Cathode material	Pt, Pt-X (X=Ru,Ni,Co)	La(Sr)MnO_3
Electrolyte	PEM (e.g. Nafion®)	ZrO_2/Y_2O_3-ceramics
Typical system power	10–100 kW	Some 100 kW
Application	e-mobility, domestic power units	Emergency power supplies

[a] Polymer electrolyte membrane fuel cell
[b] Solid oxide fuel cell

energy densities achieved in different battery systems today and expected densities in future technologies. To make connection with the volumetric energy densities for different forms of energy discussed so far in this chapter, we note that $100\,W\,h\,L^{-1}$ in Fig. 2.6 are equivalent to $3.6 \times 10^8\,J\,m^{-3}$, so that the volumetric energy density of current Li ion batteries range between 0.7×10^9 to $2 \times 10^9\,J\,m^{-3}$.

Without further discussion, we summarize in Tables 2.2 and 2.3 some properties of commonly used primary (non-rechargeable) batteries and fuel cells. As a general remark, electrochemical energy storage and conversion is probably the most dynamic field of research and development in the context of renewable and conventional energy, so that an up-to-date coverage is far beyond the scope of this book [3].

2.7 Nuclear Energy

The power generation by the use of nuclear energy is based on the different binding energies of nucleons (protons (Z) and neutrons (N)) in atomic nuclei by the strong interaction. The different binding energies for each element and some isotopes are

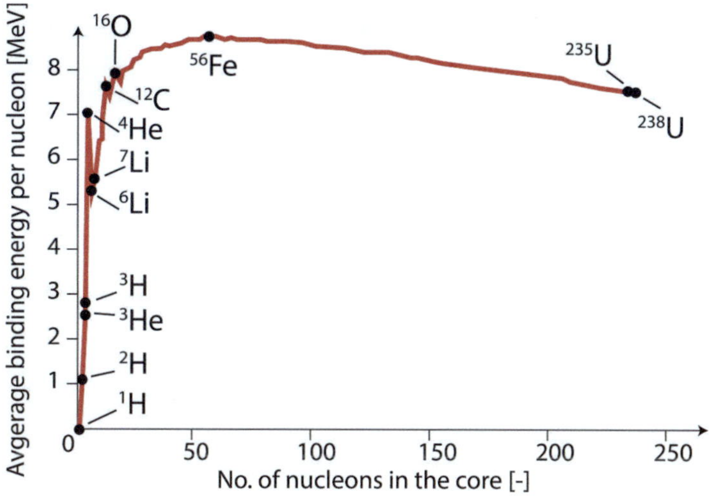

Fig. 2.7 Binding energy per nucleon as a function of the atomic number. The maximum occurs at Fe ($Z = 56$)

shown in Fig. 2.7. The maximum is reached for ^{56}Fe. For lighter elements, i.e. on the left side of this maximum, energy can be liberated by the fusion of two lighter nuclei to a heavier one, on the right side by fission of a heavy nucleus into mostly two lighter nuclei.

2.7.1 Fission

Nuclear fission, discovered by Otto Hahn and Lise Meitner in 1938, lead to the fast development of the atomic bomb as well as to nuclear power plants, of which currently 440 are in operation worldwide, producing an electric power of about 390 GW. Mainly ^{235}U is used as the fuel in these reactors. It is split by slow neutrons into two different lighter daughter nuclei and sets free a vast amount of energy. The basic process follows the reaction:

$$\mathrm{n} +^{235}\mathrm{U} \longrightarrow \mathrm{T}_1 + \mathrm{T}_2 + 2.43\,\mathrm{n} + \sim 200\,\mathrm{MeV}$$

During the fission of one uranium nucleus, on the average 2.43 neutrons are produced which can be used to maintain the fission process either in a catastrophic chain reaction or, by suitable moderation, in a constant fission rate.

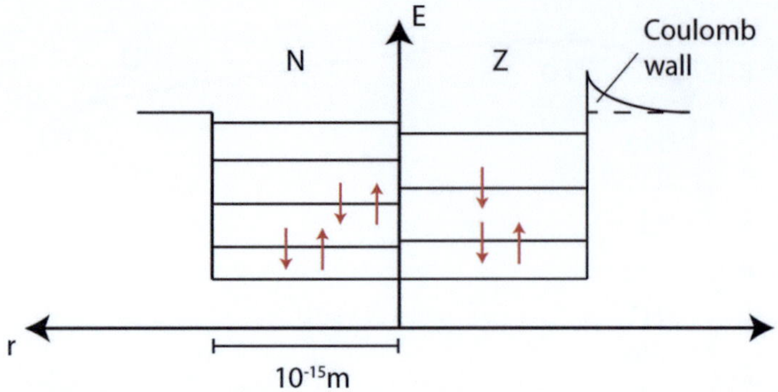

Fig. 2.8 Potential well model of a ^{7}Li nucleus. Neutrons and protons are spin $1/2$ fermions and successively occupy discrete energy levels in the nucleus until the total numbers N and Z are reached. The Coulomb interaction of charged protons gives rise to the Coulomb wall, which acts as an additional energy barrier for protons approaching the nucleus from the outside, which has to be overcome, e.g. in a fusion process

The energy density of natural uranium (99.27% ^{238}U and only 0.72% ^{235}U) is very high:

$$1\,\text{gtSCE} \mathrel{\hat{=}} 53\,\text{ktU} \quad \text{or} \quad 1\,\text{kgU} \mathrel{\hat{=}} 19\,000\,\text{kgSCE}$$

The energy density of pure ^{235}U is about

$$E/M = 7.9 \times 10^{13}\,\text{J}\,\text{kg}^{-1} \quad \text{or} \quad E/V = 1.5 \times 10^{18}\,\text{J}\,\text{m}^{-3}$$

2.7.2 Fusion

Nuclear fusion of light nuclei in hot plasmas for power generation is still under development. As shown by the energy diagram of Fig. 2.8, fusion reactions between light nuclei are driven by the gain in potential energy of the strong interaction between nucleons (neutrons and protons), but are kinetically hindered by the Coulomb repulsion between the positively charged protons in both nuclei approaching each other. This gives rise to the so-called "Coulomb wall" in the energy diagram, which has to be overcome for the fusion of charged nuclei. The outlook of cleaner nuclear energy, less radiation and shorter half-times of the radioactive reaction products as well as the large fuel (deuterium and tritium[2]) reserves is driving research since decades. The enormous requirements on the reactor materials however are making the realisa-

[2] With its short half-life time it is the bottleneck, however fusion reactors try to breed T out of Li, which is abundant.

tion extremely difficult. The focus of current research are mainly hot plasma fusion reactors with a toroidal shape. In these, the million degrees hot plasma is kept away from the walls by a strong magnetic field, which can be only achieved efficiently by superconducting magnets. The resulting vast temperature gradient from near 0 K to >10 MK shows the difficulty in finding proper materials and to advance the technology. The next experimental large scale nuclear fusion reactor ITER is currently being built in Cadarache (France) and is expected to make its first D-T fusion run in 2035, after almost 30 years of planning and construction [5]. The probably most likely reaction for fusion reactors will be the D-T fusion:

$$^{2}\mathrm{H} + ^{3}\mathrm{H} \longrightarrow ^{4}\mathrm{He} + \mathrm{n} + 17.6\,\mathrm{MeV}$$

If we use the fusion of deuterium (2.01 u[3]) and tritium (3.02 u) as an example, we find for the energy density for nuclear fusion:

$$17.6\,\mathrm{MeV} \hat{=} 2.82 \times 10^{-18}\,\mathrm{MJ}$$
$$M(D+T) = 5.03\,\mathrm{u} \hat{=} 8.34 \times 10^{-27}\,\mathrm{kg}$$
$$\frac{E}{M} = 3.4 \times 10^{9}\,\mathrm{MJkg}^{-1}$$

Fortunately, there is a very reliable and powerful fusion reactor available for us: our Sun. This will be discussed in the next chapter.

Further Reading

- A comprehensive list of gravimetric and volumetric energy densities of fuels and energy storage devices can be found on: en.wikipedia.org/energy_density. Accessed from 26 July 2022.
- Whitman, A.M.: Work and Heat. In: Thermodynamics: Basic Principles and Engineering Applications. Mechanical Engineering Series. Springer, Cham (2020). ISBN: 978-3-030-25220-5
- For a recent example of a CAES system see: caes.ppnl.gov. Accessed from 26 July 2022.
- An overview of various fly wheel applications with further references can be found on: en.wikipedia.org/wiki/Flywheel_energy_storage. Accessed from 26 July 2022.
- Statistics on global energy production: statista.com. Accessed from 26 July 2022.

[3] Atomic mass unit: $1\,\mathrm{u} = 1.66 \times 10^{-27}\,\mathrm{kg}$.

References

1. Hamann, C.H., Hamnett, A., Vielstich, W.: Electrochemistry, 2nd edn. Wiley-VCH, Weinheim (2005)978-3-527-31069-2
2. Bard, A.J., Faulkner, L.R., White, H.S.: Electrochemical Methods: Fundamentals and Applications Wiley (2022)
3. Passerini, S., et al. (Eds.): Batteries: Present and Future Energy Storage Challenges. Wiley (2020)
4. Mench, M.M.: Fuel Cell Engines. Wiley (2008)
5. Holtkamp, Norbert: An overview of the ITER project. Fusion Eng. Des. **82**, 427–434 (2007)

The Sun–Earth System 3

Abstract

The Sun is presented as the dominant energy provider for our Earth. The standard Sun model as well as energy production in the star are explained—everything in a view to the energy flux towards Earth. The second subchapter is about our planet, its structure, the energy balance on Earth and other energy and climate relevant aspects. The chapter concludes with a probable energy scenario until 2050.

3.1 The Sun

3.1.1 General Properties

Most of the energy available on Earth originates from Sun[1] as the central star of our solar system. Details of the Sun-Earth constellation are shown in Fig. 3.1. These also determine the amount of solar energy reaching the Earth. The Sun is by far the heaviest body in the system with a mass of $M_{\odot} = 2 \times 10^{30}$ kg and also the biggest with a radius of $R_{\odot} = 7 \times 10^{8}$ m (cf. Jupiter as the largest planet: $M = 1.9 \times 10^{27}$ kg at a radius of $R = 7 \times 10^{7}$ m). However it consists mainly of light atoms: 75% H, 23% He and only 2% heavier elements such as C, N and O. These are produced in the core by fusion of H and He. The average density is about $\bar{\rho}_{\odot} = 1.4$ g/cm^3, which is only a quarter of the average density of the Earth. This low value gives already reason to assume an inhomogeneous structure of the Sun, which is explained further below.

The luminosity $L_{\odot}$ of the Sun is the total power radiated by the Sun into the full solid angle $\Omega = 4\pi$:

$$\boxed{L_{\odot} = 3.82 \times 10^{26}\ \mathrm{W}} \tag{3.1.1}$$

[1] Astronomical symbol of the Sun: $\odot$.

M. Stutzmann and C. Csoklich, *The Physics of Renewable Energy*, Graduate Texts in Physics,
https://doi.org/10.1007/978-3-031-17724-8_3

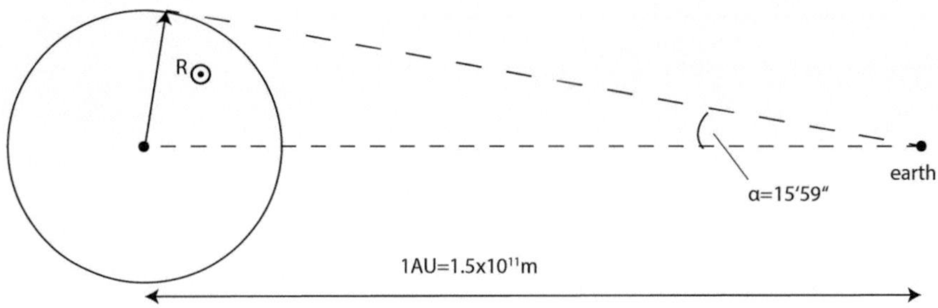

Fig. 3.1 Geometry of the Sun-Earth system. The average Sun-Earth distance defines the astronomical unit (AU). The half opening angle α of the Sun as seen from the Earth is important for concentrating solar power, cf. Sect. 6.2

$L_\odot$ is determined by measuring the **solar constant** (SC), which is the power impinging on a unit area with normal incidence just outside the Earth's atmosphere. Averaged over a year this constant is:

$$\boxed{SC = 1370\,\text{W/m}^2} \tag{3.1.2}$$

If we take the average distance between Sun and Earth with $r_{SE} = 1\,AU = 1.5 \times 10^{11}$ m into account we obtain the Sun's luminosity simply as:

$$L_\odot = 4\pi r_{SE}^2 \cdot SC$$

We can further calculate the total irradiated power over one year:

$$E_\odot = L_\odot \cdot 31536000\,\text{s} = 1.2 \times 10^{34}\,\text{J} = 3.8 \times 10^{14}\,\text{TWa} \mathrel{\hat{=}} 4 \times 10^{26}\,\text{kgCE}$$

Using $E = mc^2$, this leads to a yearly mass loss of the Sun by 1.3×10^{17} kg/a (i.e. a relative loss of 10^{-13} per year). By analyzing the spectrum of the light emitted from the Sun quantitatively, we find that the Sun radiates almost like a black body. Therefore, we can calculate the surface temperature of the Sun using the Stefan-Boltzmann law:

$$\boxed{P_{em} = eA\sigma T^4} \tag{3.1.3}$$

P_{em} stands for the total radiated power, e for the emissivity of a body ($0 \le e \le 1$, with $e = 1$ for an ideal black body), A is the emitting area (i.e. the Sun's surface), T the absolute temperature of the body and σ is the Stefan-Boltzmann constant.

$$\sigma = 5.67 \times 10^{-8}\,\text{W/m}^2\text{K}^4$$

Insertion into (3.1.3) yields:

$$P_{em} = L_\odot = 4\pi R_\odot^2 \sigma T_{eff}^4 \qquad \text{assuming } e = 1$$

$$\boxed{T_{eff} = 5780\,\mathrm{K}} \tag{3.1.4}$$

We call this an effective surface temperature, because the real temperature varies strongly from the core to the outside of the Sun.

3.1.2 Details of Proton Fusion

The generation of energy in the Sun via fusion of protons follows the overall reaction:

$$\boxed{4p^+ \rightarrow {}^4He + 2e^+ + 2\nu_e + 26.73\,\mathrm{MeV}} \tag{3.1.5}$$

The number of electron neutrinos ν_e arriving on Earth is smaller than what is expected from (3.1.5) and from the power arriving on Earth. The reason for this effect are neutrino oscillations: electron, muon and tau neutrinos periodically transform during their lifetime into each other (i.e. change their flavor). This quantum effect indicates that neutrinos have mass and describe hereby a theory beyond the Standard Model [1,2].

This overall reaction can be achieved via different reaction pathways. The most direct one starts with the fusion of two protons to form a deuteron (np^+):

$$\mathrm{p}^+ + \mathrm{p}^+ \longrightarrow \mathrm{d} + \mathrm{e}^+ + \nu_e + 0.42\,\mathrm{MeV} \tag{3.1.6}$$

Due to the high temperatures of several million K necessary to achieve fusion, all atoms in the Sun are completely ionized, so that hydrogen exists in the form of a plasma formed by free protons and free electrons. The resulting positron is needed to conserve charge, and the electron neutrino ensures conservation of the lepton number. After its emission, the positron annihilates immediately with a free electron via the emission of two γ-photons.

In the next step, a deuteron fuses with a third proton to form ^{3}He:

$$\mathrm{d} + \mathrm{p}^+ \longrightarrow {}^3\mathrm{He} + \gamma + 5.49\,\mathrm{MeV} \tag{3.1.7}$$

After the formation of ^{3}He, there are three possible reactions to produce ^{4}He, starting at different temperatures:

1. Dominating at $\mathrm{T} = 10 - 14 \times 10^6$ K:

$$^3\mathrm{He} + {}^3\mathrm{He} \longrightarrow {}^4\mathrm{He} + 2\,\mathrm{p}^+ + 12.86\,\mathrm{MeV}$$

2. Dominating at $T = 14 - 23 \times 10^6$ K

$$^3\text{He} + \text{p}^+ \longrightarrow {}^4\text{He} + \text{e}^+ + \nu_\text{e} + 18.77\ \text{MeV}$$

3. Dominating at $T > 23 \times 10^6$ K (with ^{4}He as a catalyst for higher production of ^{4}He):

$$\begin{aligned} {}^3\text{He} + {}^4\text{He} &\longrightarrow {}^7\text{Be} + \gamma + 1.59\ \text{MeV} \\ {}^7\text{Be} + \text{e}^- &\longrightarrow {}^7\text{Li} + \nu_\text{e} \\ {}^7\text{Li} + \text{p}^+ &\longrightarrow 2\,{}^4\text{He} + 17{,}35\ \text{MeV} \\ \text{and:} & \\ {}^7\text{Be} + \text{p}^+ &\longrightarrow {}^8\text{B} + \gamma + 0.14\ \text{MeV} \\ {}^8\text{B} &\longrightarrow {}^8\text{Be}^* + \text{e}^+ + \nu_\text{e} \\ {}^8\text{Be}^* &\longrightarrow 2\,{}^4\text{He} \end{aligned}$$

3.1.2.1 CNO-Cycle (Bethe-Weizsäcker Cycle)

In stars with more than 1.3–1.5 $M_\odot$ (core temperatures $T_{core} \geq 20$ MK) the CNO- or Bethe-Weizsäcker-cycle dominates. This fusion chain is catalysed by the heavier nuclei C, N, and O and also produces ^{4}He. In our Sun this reaction contributes about 1.6% to the overall solar energy production. In the reaction chain intermediate, unstable nuclei are involved, i.e. ^{13}N and ^{15}O, which decay by positron emission:

- $^{12}\text{C} + \text{p}^+ \longrightarrow {}^{13}\text{N} + \gamma$
  $^{13}\text{N} \longrightarrow {}^{13}\text{C} + \text{e}^+ + \nu_\text{e}$ (7 min)
- $^{13}\text{C} + \text{p}^+ \longrightarrow {}^{14}\text{N} + \gamma$
- $^{14}\text{N} + \text{p}^+ \longrightarrow {}^{15}\text{O} + \gamma$
  $^{15}\text{O} \longrightarrow {}^{15}\text{N} + \text{e}^+ + \nu_\text{c}$ (82 sec)
- $^{15}\text{N} + \text{p}^+ \rightarrow {}^{12}\text{C} + {}^4\text{He}$

With the overall reaction being again:

$$4\,\text{p}^+ \longrightarrow {}^4\text{He} + 2\,\text{e}^+ + 2\nu_e \tag{3.1.8}$$

Our Sun is currently only producing elements up to carbon (C, Z = 8). Elements heavier than C can be created in later stages by dying stars depending on their initial mass. Stars can create elements up to iron (Fe, Z = 56) by fusion. Beyond that, the energy balance becomes negative (cf. Sect. 2.7). The production of heavier elements requires far higher energies which are only found in novae, supernovae and similar events.

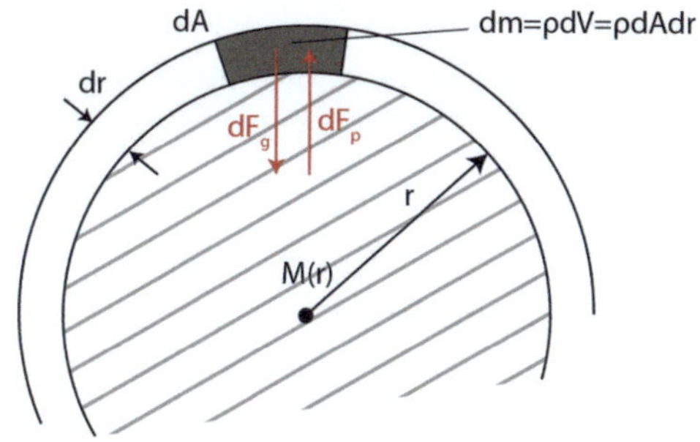

Fig. 3.2 Equilibrium of pressure and gravity in a star. $M(r)$ is the total mass in a sphere with radius r

3.1.3 Shell Model

A simple model of the Sun can be derived by considering the equilibrium of forces and energy production in a stationary star[2] and assuming an ideal gas for the particles in the star. This model is called "shell model", since the following considerations lead to a shell structure. This model considers the interrelations between the density ρ, the pressure p, the absolute temperature T and the energy flux Φ as a function of the distance r from the star center. It is based on the following three relations:

1. **The Ideal Gas Equation:**

$$\boxed{p(r) = \frac{k_B T(r)}{\bar{m}} \rho(r)} \tag{3.1.9}$$

   Here, $\bar{m}$ is the effective mass per particle $(1/2(m_p + m_e) \approx 1/2m_p)$.

2. **Equilibrium of Pressure and Gravitation:**
   A mass element dm in Fig. 3.2 is given by

$$\mathrm{d}m = \rho\,\mathrm{d}V = \rho\,\mathrm{d}A\,\mathrm{d}r$$

   Each of these mass elements is subjected to the gravitational force, as well as the pressure force caused by the core's fusion reaction:

$$\mathrm{d}F_g = G\frac{M(r)}{r^2}\,\mathrm{d}m = G\frac{M(r)}{r^2}\rho(r)\,\mathrm{d}A\,\mathrm{d}r \tag{3.1.10}$$

$$\mathrm{d}F_p = [p(r+dr) - p(r)]\,\mathrm{d}A = -\frac{\mathrm{d}p(r)}{\mathrm{d}r}\,\mathrm{d}r\,\mathrm{d}A \tag{3.1.11}$$

---

[2] Of course this model cannot explain the time evolution of a star from formation to death. The interested reader is referred to the corresponding literature.

Here, G is the gravitational constant and $M(r)$ is the total mass inside r. For a stationary star, these two forces (3.1.10 and 3.1.11) must be equal and opposite:

$$\boxed{\mathrm{d}F_g = \mathrm{d}F_p \Rightarrow \frac{\mathrm{d}p}{\mathrm{d}r} = -G\frac{M(r)}{r^2}\rho(r)} \tag{3.1.12}$$

3. **Energy Production:**
The differential energy e_{em} produced in a shell dr with radius r is:

$$e_{em}(r) = \varepsilon\,\mathrm{d}M(r) = \varepsilon 4\pi r^2 \rho(r)\,\mathrm{d}r \tag{3.1.13}$$

ε is the energy produced per unit mass. The released energy can be described as an energy flux $\Phi(r)$ through the surface of a sphere with radius r:

$$\mathrm{d}\Phi = \Phi(r + \mathrm{d}r) - \Phi(r) = \varepsilon\,\mathrm{d}M(r)$$

This can be rewritten with (3.1.13) as:

$$\boxed{\frac{\mathrm{d}\Phi}{\mathrm{d}r} = 4\pi r^2 \varepsilon\rho(r)} \tag{3.1.14}$$

These coupled equations (3.1.9, 3.1.12 and 3.1.14) can be solved with the known boundary conditions at the surface of the Sun:

$$M(R_\odot) = M_\odot = 2 \times 10^{30}\,\mathrm{kg} \tag{3.1.15}$$

$$T(R_\odot) = T_{eff} = 5780\,\mathrm{K} \tag{3.1.16}$$

$$\Phi(R_\odot) = L_\odot = 3.82 \times 10^{26}\,\mathrm{W} \tag{3.1.17}$$

$$p(R_\odot) = 0 \tag{3.1.18}$$

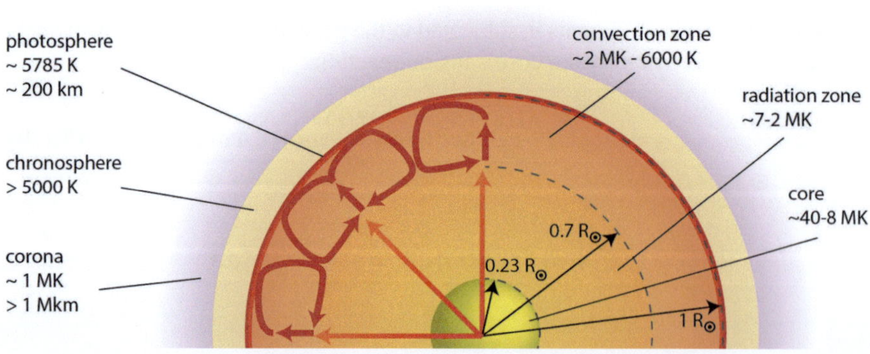

Fig. 3.3 Shell model of the Sun. See text for more details

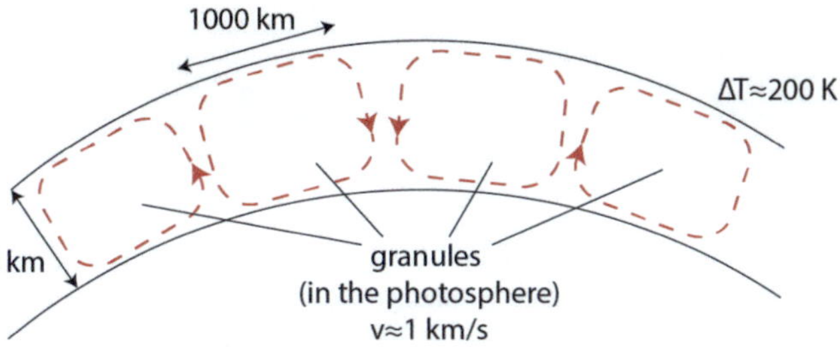

Fig. 3.4 Convection cells in the photosphere

This leads to the following shell model of the sun (Figs. 3.3 and 3.4):

- 90% of the energy is produced in the core with radius $r \leq 0.23 R_{\odot}$. The core temperature is 15 MK, the core density $\rho_{core} = 100\,\text{g/cm}^3 \approx 70\rho_{average}$.
- The energy generated in the core is transported by γ-radiation up to 0.7 $R_{\odot}$. In the outer shell energy is transported by convection cells.
- The visible radiation emitted by the Sun originates from the so-called *photosphere*, a thin surface layer of about 200 km, where the temperature drops from $T \approx$ 8000 K to $T \approx$ 4500 K. The photosphere also has a granular character due to convection cells.
- Beyond the photosphere exists the chromosphere with a thickness of about 10 000 km. It consists of hot particles able to leave the photosphere according to the barometric height formula. The emitted radiation consists of specific atomic spectra (H, He, Mg, Fe).
- Further outside is the corona, consisting of very hot particles (T≈1 MK), extending about 1×10^6 km. The most energetic particles can leave the gravitational field of the Sun as the solar wind. This gives rise to an additional mass loss of $\approx 10^{-13} M_{\odot}$ per year
- In addition the Sun exhibits dynamical events of statistical nature:

  Flares: short eruptions (minutes to hours)

  Protuberances: fast particles remaining up to weeks in the external magnetic field of the Sun

  Sunspots: local disturbances in the magnetic field. The magnetic field lines reach through the photosphere and hinder thereby heat transport in these regions. Thus they are colder and appear darker (Fig. 3.5). Sunspots indicate increased solar activity with a more disordered magnetic field pattern.[3] Due to different rotation speeds of the different shells, the number of sunspots varies with a period of around 11 years (9–15 years, cf. Fig. 3.6).

[3] First observations date back to the Antique. Among others, Galileo kept such a sunspot record for a couple of years. Since the 19th century different records were collected and thus a long data base was created.

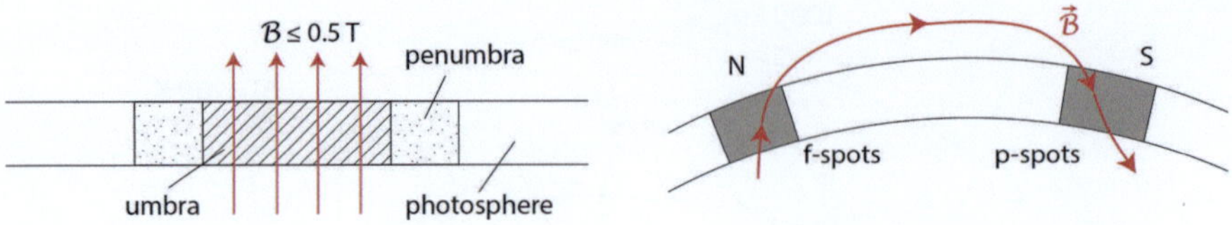

Fig. 3.5 Typical structure of sunspots. Left: Umbra with surrounding penumbra. Right: Sunspots appear mostly in pairs with common magnetic field lines and in combination with flares

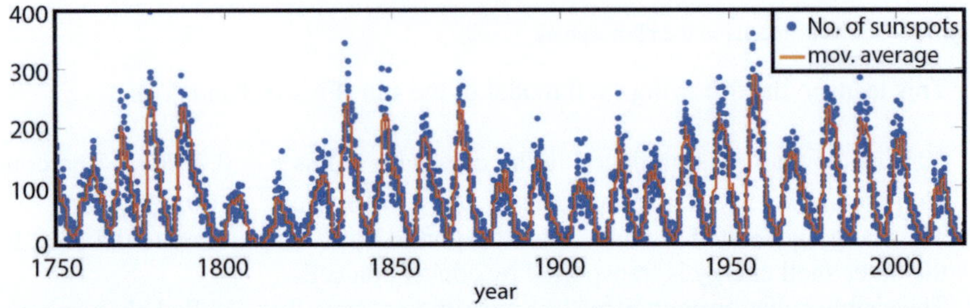

Fig. 3.6 Sunspot cycles of the past 400 years. Blue dots indicate the monthly average of sunspots, the orange line a moving average across 10 data points. *Source* WDC-SILSO, Royal Observatory of Belgium, Brussels [3]

3.2 The Earth

3.2.1 General Properties

With a radius of only $R_♁ = 6.4 \times 10^6$ m and a mass of $M_♁ = 5.974 \times 10^{24}$ kg, the Earth[4] is the third planet by distance from the Sun. Using a simple geometrical ansatz, we can approximate the total incoming power from the Sun:

$$P_♁ = (1 - R)\pi R_♁^2 SC \tag{3.2.1}$$

$\pi R_♁^2$ is the surface area projection towards the Sun. The additional term (1-R) with the reflectivity R takes into account that not all incoming energy is absorbed. The Earth's surface reflects a certain percentage, varying significantly across our planet. Typical values of R are listed in Table 3.1.[5]

[4] Astronomical symbol of the Earth: ♁.

[5] The measure for the reflection is called reflectivity R. Together with transmittance T and absorption A balance must hold: $1 = T + A + R$.

Table 3.1 Typical reflectivities of different surfaces on Earth

Surface	Reflectivity
Clouds	0.2–0.7
Water (at the equator, normal incidence)	0.05
Water (northern hemisphere)	0.25
Snow	0.3–0.7
Green vegetation	0.1–0.2
Deserts	0.3

Overall, this leads to a global average reflectivity of $R \approx 30\%$. Together with the SC we then obtain for the total incoming power[6]:

$$P_{\text{♁}} \approx 1.2 \times 10^{17}\,\text{W} = 1.2 \times 10^{5}\,\text{TW}$$

From this, we can calculate the Earth's theoretical average surface temperature T_E, if we set up the energy balance between incoming solar radiation and emitted thermal radiation of the Earth. In a first approximation we assume that the Earth is a black body as well with $e = 1$:

$$P_{\text{♁}} = (1 - R)\pi \cdot R_{\text{♁}}^2 \cdot SC = 4\pi \cdot R_{\text{♁}}^2 \cdot \sigma \cdot T_E^4$$

$$\Rightarrow T_E = \left(\frac{(1 - R)SC}{4\sigma}\right)^{1/4} \approx 254\,\text{K}$$

Thus, the average equilibrium surface temperature of the Earth under these assumptions would be only $T_E \approx -19\,°\text{C}$. However, the observed average temperature today is $\overline{T}_{\text{♁}} = 288\,\text{K} \approx 15\,°\text{C}$. This important difference is caused by the greenhouse effect: without the additional reflectivity of our atmosphere also for the outgoing radiation from the Earth's surface, temperatures would have been most likely too cold for life based on liquid water.

The wavelength λ_{max} of the maximum of the emitted black body radiation can be calculated with the Wien-law based on the experimentally observed spectra of the incoming and emitted radiation (cf. Fig. 3.7):

$$\boxed{\lambda_{max} = \frac{\text{hc}}{5\text{k}_\text{B}\text{T}} = \frac{2898}{\text{T}}\,\mu\text{m K}} \tag{3.2.2}$$

$$\Rightarrow \lambda_{max,\odot} = 490\,\text{nm(green-blue)}$$

$$\Rightarrow \lambda_{max,\text{♁}} = 12\,\mu\text{m(infra-red)}$$

[6] In addition to the black body radiation of the Sun, the Earth receives a weak spectrum of γ-rays from the corona ($\lambda \approx 0.1$nm).

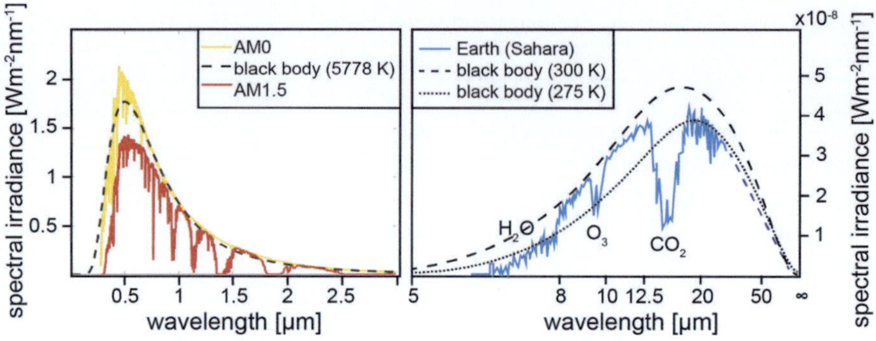

Fig. 3.7 Left: Quantitative comparison of the emitted radiation of the Sun arriving on Earth before (yellow) and after passing Earth's atmosphere (red) [4]. Right: Earth's infrared emission spectrum with characteristic absorption peaks of mainly H_2O, O_3, and CO_2. Note the orders of magnitude difference between the left and right y-axis. *Source* ASTM G-173 AM1.5 Reference spectra, Earth thermal infrared emission spectrum as captured over the Sahara desert by the Nimbus 4 satellite [5]

To understand and model the very important modulation of the radiation emitted by the Sun and the Earth when passing through our atmosphere, several relevant physical processes have to be taken into account:

- The infrared spectrum emitted by the Earth is modulated via absorption and reflection by strong vibrational absorption bands mainly of H_2O and CO_2 in the wavelength range between $\lambda = 5\,\mu\text{m}\,to\,30\,\mu\text{m}$. The current average concentrations of H_2O and CO_2 are 1600 ppm and 400 ppm, respectively. This provides the physical basis for the notorious greenhouse effect, which however is further complicated by the existence of (partly fragile) feedback loops. Thus, the increase in CO_2 due to burning of fossil fuels leads to in increase in the surface temperature of the Earth, resulting in an additional increase of water in the atmosphere due to increased evaporation. The increased water concentration will further strengthen the greenhouse effect (positive feedback), but also leads to the formation of more clouds, increasing the reflectivity for the incoming solar radiation (negative feedback).
- The solar radiation arriving on the Earth surface is modulated by electronic absorption bands especially of O_3 (ozone), O_2, H_2O, and CO_2. Ozone is mainly responsible for absorption in the ultraviolet region of the solar spectrum (explaining the danger related to the ozone hole), atmospheric oxygen mainly absorbs in the visible region, and water as well as carbon dioxide absorb solar radiation in the red and near-infrared region. The integral absorption of Sun light via the atmosphere is described by the quantity air mass (AM), which takes into account the different effective length that sunlight travels through the atmosphere, depending on the geographical location:
  - $\text{AM} = 1/\sin\gamma$, where γ is the zenith angle
  - AM 1: Sun is in zenith (at the equator)
  - AM 1.5: $\approx$ Munich (with a maximum intensity of $1\,\text{kW/m}^2 = 100\,\text{mW/cm}^2$)

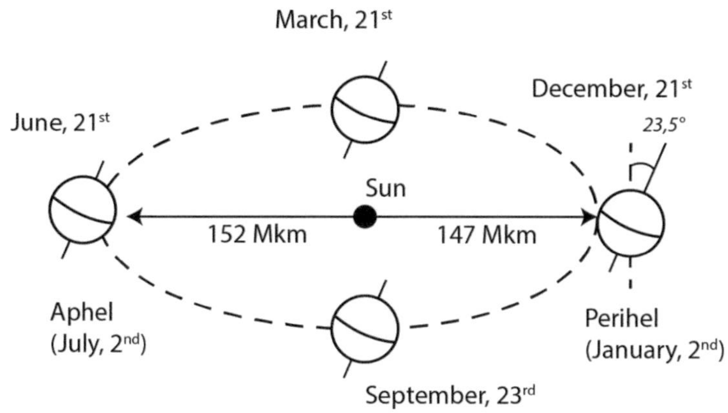

Fig. 3.8 Schematic view of the elliptical orbit of the Earth around the Sun

- The geothermal energy flux through the surface of the Earth (cf. Chap. 6) is less than 0.01% of the impinging solar energy flux and can be neglected in these considerations.

Since the Earth orbits around the Sun on an elliptical path, the distance between Sun and Earth exhibits a periodic modulation throughout the year. This causes also varying values of the solar constant and, thereby, the incoming flux of energy:

- SC = 1415 W/m^2 in January (Perihel, Earth-Sun distance 147 Mkm)
- SC = 1325 W/m^2 in July (Aphel, Earth-Sun distance 152 Mkm)

In addition, the rotational axis of the Earth is inclined by 23.5° with respect to the orbit plane of the Earth (ecliptic).[7] This gives rise to the different seasons in the northern and southern hemispheres, leading to locally even more pronounced yearly variations of the incoming solar radiation (Fig. 3.8).

In addition to the dominant annual variation of solar irradiation due to the elliptical Earth orbit and the rotational axis inclination, the solar luminosity $L_\odot$ varies as well with time. On the one hand there exist statistical variations of the luminosity by about 0.5% (causing a corresponding variation of the surface temperature of the Earth by about $\Delta T_{♁} \approx 0.4\,°\mathrm{C}$). On the other hand, there are periodical variations by a similar amount correlated with the number of sunspots. As discussed above, the number of sunspots varies between 0 and about 300 spots with a period of about 11 years. There also are indications of an additional super-period of about 100 years (see Fig. 3.6). Since many sunspots correspond to an increased activity of the Sun, this leads to

[7] The currently accepted reason for this inclination of the rotational axis of the Earth with respect to its orbit plane around the Sun is the collision of the Earth with a Mars-like cosmic object, which caused the shift of the axis and the simultaneous formation of our Moon from the ejected debris.

additional long-term variations of the solar luminosity and the surface temperature of the Earth.

A still debated correlation is that between the so-called *Maunder minimum*, an abnormal reduction of observed sun spots during the years 1645 and 1715, and the "little ice age" in Europe and Northern America in about the same time range, which lead to a severe famine catastrophe in Europe. Alternative or additional explanations may come from enhanced volcanic activity prior and during this time.

Figure 3.9 shows a schematic summary of the main global energy fluxes caused by solar irradiation on the Earth. The main fluxes are linked to optical processes (absorption, reflection and thermal radiation) and thermo-mechanical processes (convection, evaporation). Compared to those, geothermal, gravitational or biomass-related fluxes are almost negligible. Yet, the latter have a very large influence on our daily life as well.

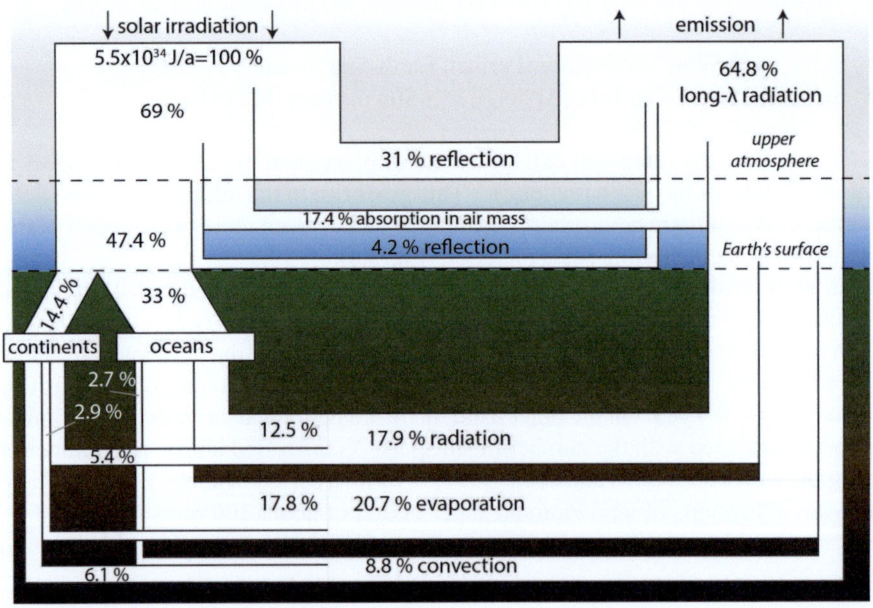

Fig. 3.9 Schematic view of the energy fluxes on Earth

3.3 A Possible Energy Scenario Until 2050

In the following we give a personal view of the requirements, boundary conditions and possible solutions to provide our planet with sustainable energy by the year 2050, based on the present situation in 2020. As a word of caution: this or similar long-term scenarios always should be viewed with critical eyes, because political, scientific, technological, ecological and societal aspects are likely to change significantly over such long time spans.

We assume a continuous increase of the world population up to 10 billion people by 2050. Moreover, we estimate an average consumption of 2 kW/human being, including 1 kW for food production. Note that this means that most developed countries will have to save 60–90% of their current energy consumption, which already would constitute a tremendous task. With this we require a continuous necessary power "to run the planet" of 20 **TW**.

At the moment, we still have sufficient fossil fuel reserves to maintain such an energy production for more than thousand years. Current estimates predict sufficient oil supplies for about 50–150 years, gas for 200–600 years and coal for more than 2000 years. However, if we want to keep global warming due to the resulting emission of CO_2 below an average temperature rise of 2 °C, we have to stop the use of fossil fuels as soon as possible, reaching zero emission globally by 2050.

Possible solutions for this energy problem are very limited:

- Increased use of nuclear power (fission only, since fusion will very likely remain on an experimental scale until 2050).
- Active removal of CO_2 from the atmosphere by carbon capture and storage (CCS) and/or artificial photosynthesis.
- Global geo-engineering to counteract the greenhouse effect by climate cooling.
- Ramping up the use of renewable energy to the required level of 20 TW by 2050.

So let us have a critical look at these four possibilities one by one.

(1) Nuclear Power
To provide the required 20 TW, about 20 000 new nuclear power plants with a typical power of 1 GW would have to be built. This means that until 2050 every day at least one new plant would have to go online. In reality, today less than 100 new nuclear reactors are planned or under construction, with usually more than ten years required for completion. Moreover, the known uranium reserves for several thousand nuclear plants would only last for a few decades, so that the risky breeder technology would have to be used worldwide on a large scale.

(2) Carbon Capture and Storage and Artificial Photosynthesis
The global annual CO_2 emission by mankind has increased from about 6 Gt per year (Gt/a) in 1950 to about 40 Gt/a today. Of these, about 10% alone are produced by the metabolism of 8 billion humans and their domestic animals and, thus, are almost unavoidable. In comparison, the net uptake of atmospheric CO_2 by the oceans only amounts to about 2 Gt/a and already has led to a noticeable and dangerous

acidification of the sea water. Storing CO_2 in solid form in deep cold water at high pressures is physically possible, but has never been tested on a large scale for a sufficiently long time. The same is the case for CO_2 storage underground on land and usually meets great public concern even for small CCS pilot plants. Anyhow, given the large amount of CO_2 that would need to be stored annually (about ten times the worldwide concrete and cement production of 5 Gt/a) and the related energy consumption, avoiding CO_2 emission at all would be the much better alternative. In this respect, also artificial photosynthesis, which aims at removing CO_2 from the atmosphere and transform it with the help of sunlight into fuels like methane or methanol is a very interesting approach (see Chaps. 7 and 8). However, even after decades of intensive research efforts, efficient large scale and long-term stable systems made from sustainable materials will not be available in the foreseeable future.

(3) Climate Cooling by Geo-Engineering

Geo-engineering aims at the global manipulation of the climate by large scale removal of CO_2 from the atmosphere as discussed above, or at a global increase of the Earth's reflectivity for a reduction of incoming sunlight. Methods for the latter approach currently discussed are mirrors placed in an Earth orbit, distribution of small reflecting particles like sulfur dioxide in the upper atmosphere, or increasing the reflectivity of the Earth surface by white paint or mirrors. That such approaches indeed can significantly lower the global temperature was last demonstrated by the eruption of the Pinatubo volcano in 1991, which caused a temperature decrease of about 0.5°C. However, it is still not possible to predict the local consequences and risks of geo-engineering sufficiently well, and many scientists fear possible conflicts between different countries or even the use of geo-engineering as a weapon. [8]

(4) Ramping up the Use of Renewable Energy Sources

Probably the most plannable, controllable, sustainable and, therefore, most likely solution will be the replacement of fossil fuel usage by CO_2-neutral renewable energy sources. Before we discuss this in detail in the following chapters, we summarize here briefly our view of the potential, limitations and pending problems of renewable energy sources.

Energy from Rivers, Waves and Tides

Running water in streams and rivers is estimated to represent a total power of 5–6 TW. Of these, however, only around 2 TW can be harvested in an ecologically and economically viable manner. Especially in developed countries, most of these viable resources are already being used. Random small scale waves in the oceans and periodic large scale tidal waves are estimated to be able to contribute an additional power of 2 TW.

[8] As a side note, allowing CO_2 levels to further increase also has another consequence: beyond levels of 1000 ppm the intellectual performance and capabilities of humans are noticeably decreased.

Wind Energy
The global winds on-shore and off-shore are estimated to be able to realistically contribute a total power of 4 TW to 5 TW. To fully harvest this source, about one million wind engines with a power capacity of 5 MW each would be necessary. This translates into 100 of such wind engines which would need to be constructed every day until 2050, for a yearly increase of wind power by 180 GW/a. For comparison, the currently available wind power is increasing by about 60 GW/a, so "just" a factor of three off.

Photosynthesis and Biomass
Although the conversion of solar energy via photosynthesis by plants, algae and bacteria in the past and at present forms the basis of our food and energy supply, the low average efficiency of natural photosynthesis of only 0.3% limits its potential upscaling. For the production of 20 TW, about 30% of the available continental area would be required. In addition, a sufficient supply of water would be required in these areas. Conservative estimates of the potential contribution of biomass to the pool of renewable energy limit it to 5 TW.

Direct Use of Sunlight by Photovoltaics, Artificial Photosynthesis and Solar Thermal Power
According to Chap. 2, the constant power provided by solar radiation on Earth is 120000 TW, with a density of about 1 kW m^{-2}. So, by transforming solar irradiation into electricity with a 20% efficient solar cell, the entire power presently consumed by the USA of about 3 TW could be produced by covering 2% of the US area with such solar cells. This corresponds roughly to the area covered by streets, roads and highways. To reach a power capacity of 20 TW by 2050, about one million solar modules with a power output of 1 kW would have to be installed worldwide every day, an annual increase by 365 GW/a. This does not seem impossible given the current global photovoltaic annual growth rate of 60 GW/a.

To summarize, a sustainable energy supply of mankind without further detrimental damages to our ecosphere is possible, even on the time scale of the three decades remaining until 2050. But it will require serious and concerted efforts in (i) saving energy especially in developed countries, (ii) replacing conventional energy sources by an intelligent mix of renewable energy worldwide, and (iii) developing distributed energy storage capacities on the TW scale to buffer the daily and seasonal fluctuations of the solar energy available at a given place and time.

Further Reading

- Guenther, D.B. et al.: Standard solar model. Astr. Phys. J. Part 1, **387**, 372–393 (1992).
- Currently, the NASA's *Parker Solar Probe* investigates the surface of the Sun and will be the closest satellite to the Sun ever launched. More information and images can be found on the webpage. www.nasa.gov/content/goddard/parker-solar-probe. Accessed from July 2022.

References

1. Eidelman, S. et al.: Review of particle physics. Phys. Lett. **B592**(1–4) (2004)
2. Ahn, M.H., et al.: Measurement of neutrino oscillation by the K2K experiment. Phys. Rev. D **74**, 072003 (2006)
3. Sunspot data from SILSO Royal Observatory of Belgium, Brussels. www.sidc.be/silso. Accessed from 26 July 2022
4. ASTM G-173 AM1.5 Reference spectre, available at NREL's website: https://nrel.gov/grid/solar-resource/spectra-am1.5.html. Accessed from 27 July 2022
5. Hanel, R.A., Conrath, B.J.: Thermal emission spectra of the Earth and atmosphere from the nimbus 4 Michelson interferometer experiment. Nature **228**, 143–145 (1970)

Energy from Waves, Tides and Osmosis 4

Abstract

This chapter deals especially with wave power and osmosis power plants. The physics of wave generation and propagation as well as the wave's energy content are explained. As a special form of waves, tidal waves are discussed in more detail. Principles of osmosis and how it can be used for power generation are also briefly explained.

4.1 Wave Energy

Water waves are created by friction and lift forces of wind streaming over a water surface. They are an example of the general Kelvin-Helmholtz instability at the boundary of two media moving with different velocities. By this instability, random fluctuations at the interface are amplified. An initial wave crest is further lifted up according to the Bernoulli equation, because the wind velocity v_{wind} increases at the wave crest and, as a result, the air pressure p_{air} is reduced there:

$$p_{air} + \frac{1}{2}\rho_{air} v_{wind}^2 = \text{const.}$$

In addition, the wind can push a wave forward in the wind direction because of friction and impact pressure, as sketched in Fig. 4.1.

For a non-progressing wave in deep water (meaning that the total water height h is at least larger than a quarter of the wave length λ_w), the water molecules move on closed circles, with radii R decreasing exponentially with depth d (deep water waves). If the wave is also progressing with mass transport along the water surface, the speed of progression adds to the circular velocity and gives a forward spiraling orbital motion of the molecules. The circular motion in non-progressing deep water waves is also perturbed when the total water height h becomes less than $\lambda_w/4$, as shown in Fig. 4.2:

M. Stutzmann and C. Csoklich, *The Physics of Renewable Energy*,
Graduate Texts in Physics,
https://doi.org/10.1007/978-3-031-17724-8_4

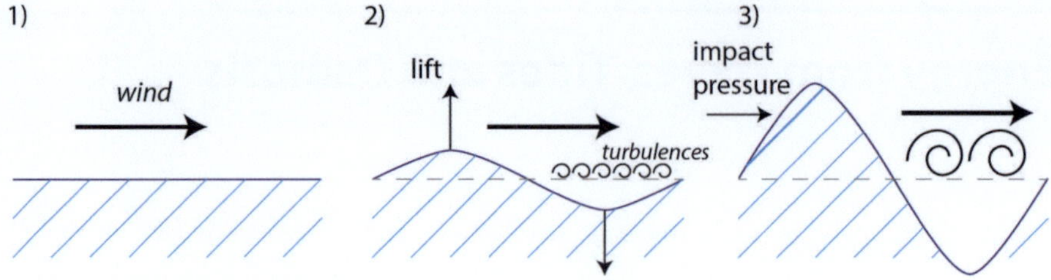

Fig. 4.1 Illustration of the interaction of wind with the water surface, leading to the creation of waves (and turbulences)

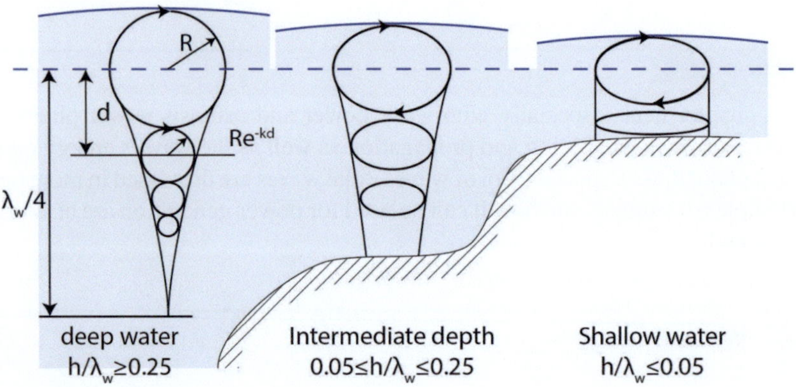

Fig. 4.2 Local orbits of molecules in deep and shallow water waves (λ_w is the wavelength, h the water height)

4.1.1 Deep Water Waves

The propagation of waves in deep water (water depth $h > \lambda/4$) follows a simple consideration: the wave does not "feel" the ground, i.e. the interaction of the water motion with the resting bottom of the sea can be neglected. Then, the local circular trajectories at the surface occur with a velocity $v_{circle} = 2\pi\ R \cdot f_w$, where f_w is the frequency of the wave. Although the wave movement has a finite phase velocity v_w, the single molecules do not propagate averaged over time (cf. Fig. 4.3).

This local circular movement causes a periodic transformation between potential and kinetic energy of the molecules. At the surface, the molecules are falling by a height of $\Delta h = 2R$, with respect to the equilibrium surface. The potential energy has its maximum at the wave crest, whereas the velocity $v_{top} = v_w - v_{circle}$ reaches a minimum. At the wave trough, the potential energy is minimal, whereas the velocity $v_{bottom} = v_w + v_{circle}$ has its maximum. The resulting changes in kinetic and potential energy are equal, from which the dispersion relation $v_w\ (\lambda_w)$ can be derived:

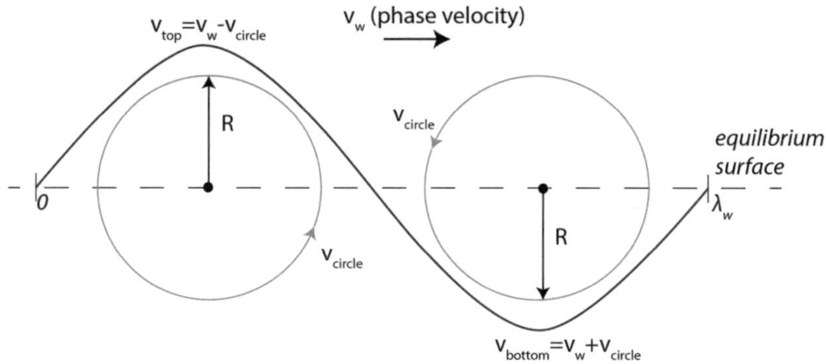

Fig. 4.3 Motion of water particles during a wave transition

$$\Delta E_{kin} = \frac{1}{2}m\left(v_{bottom}^2 - v_{top}^2\right) = \frac{1}{2}m\,(4v_w\,v_{circle}) = 2m\,v_w 2\pi\,R\,f_w$$
$$\Delta E_{pot} = mg \cdot 2R$$
$$\Delta E_{kin} = \Delta E_{pot}; \qquad \Rightarrow 2\pi\,v_w f_w = g \qquad \text{and:}$$

$$\boxed{v_w = \frac{g}{2\pi\,f_w}} \tag{4.1.1}$$

With $f_w = v_w/\lambda_w$ this gives:

$$\boxed{v_w = \sqrt{\frac{g\lambda_w}{2\pi}}} \tag{4.1.2}$$

Thus, deep water waves have a pronounced dispersion: long waves move faster than short waves. Since the dispersion relation is only determined by the gravitational acceleration of the Earth, such waves are also called gravitational waves.

The dispersion of deep water waves can be seen by the following phase velocities for different wavelengths:

$$\lambda_w = 10m \Rightarrow v_w = 4 \text{ m/s}$$
$$\lambda_w = 100m \Rightarrow v_w = 12.5 \text{ m/s}$$
$$\lambda_w = 1000m \Rightarrow v_w = 40 \text{ m/s}$$

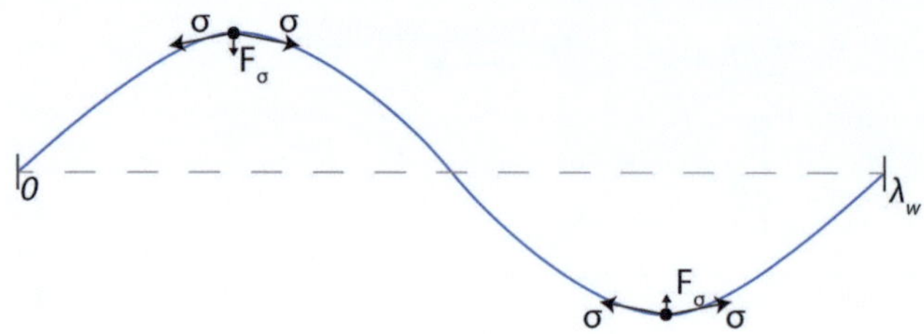

Fig. 4.4 Additional restoring force caused by the surface tension σ, dominating the wave velocity for small wavelengths

In addition to gravity the velocity of water waves is also influenced by surface tension and by a finite depth of the water [1]. The surface tension σ[1] adds a restoring force to the wave as seen in Fig. 4.4. Every deviation from a flat surface increases the surface area of a uniform flat water surface at rest. This increases the surface energy, which then tries to reach a minimal value again. The influence of surface tension modifies the dispersion relation mainly at very small wavelengths:

$$\boxed{v_w = \sqrt{\frac{g\lambda_w}{2\pi} + \frac{2\pi\sigma}{\rho\lambda_w}}} \tag{4.1.3}$$

This gives rise to two regimes of increasing phase velocity with either increasing or decreasing wavelength:

Capillary waves ($\lambda_w < 1\,\mathrm{cm}$)

$$v_w \propto \sqrt{1/\lambda_w}$$

Gravitational waves ($\lambda_w > 10\,\mathrm{cm}$)

$$v_w \propto \sqrt{\lambda_w}$$

4.1.2 Shallow Water Waves

A finite water depth d influences the wave propagation noticeably for $d \le \lambda_w/4$. Then the wave interacts with the ground, which leads to a more elliptical trajectory of the molecules as shown in Fig. 4.2. The velocity of a shallow water wave is given by:

$$\boxed{v_w = \sqrt{\frac{g\lambda_w}{2\pi}\tanh\left(\frac{2\pi d}{\lambda_w}\right)}} \tag{4.1.4}$$

[1] (with $\sigma = 7 \times 10^{-2}$ N/m, $\rho = 1 \times 10^3$ kg/m^3 for water).

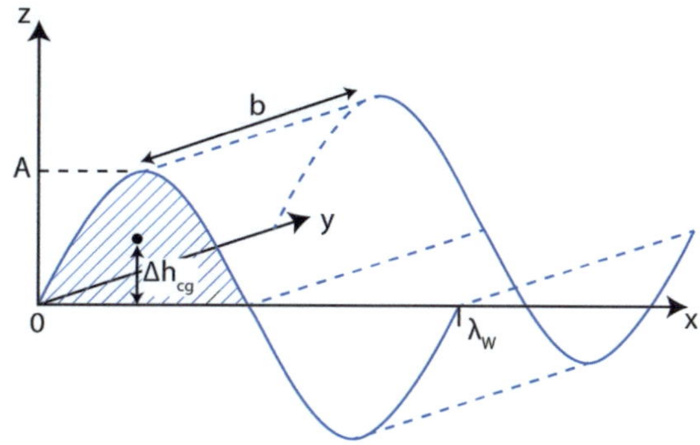

Fig. 4.5 Illustration of wave parameters involved in the energy calculation. Δh_{cg} indicates the height of the center of gravity of the wave

For $d << \lambda_w$ ($\tanh(x) \to x$), this can be approximated by $v_w = \sqrt{g \cdot d}$, so that shallow water waves have no dispersion (no dependence of v_w on λ_w).

This is the reason for surf or tsunami waves: long deep water waves with a high wave velocity approach the shore, where their behaviour is influenced more and more by the decreasing depths $d < \lambda_w/4$ and they "become" shallow water waves with decreasing wave velocity. Energy conservation leads then to an increase of the wave amplitude (i.e. potential energy) when reaching the coast.

Finally we will estimate the energy content of a wave. Let us assume a wave front with a lateral extension of b, an amplitude A, and wavelength λ_w (Fig. 4.5). For a sinusoidal wave the upper half of the wave contains a water mass M of:

$$M = \rho V = \rho b A \int_0^{\lambda_w/2} \sin\left(\frac{2\pi x}{\lambda_w}\right) dx = \rho \cdot b \cdot A \cdot \frac{\lambda_w}{\pi}$$

For $b = 1$ km, $A = 2.5$ m, and a wavelength $\lambda_w = 50$ m, this results in a mass of about $M \approx 40\,000$ t. The position of the centre of gravity Δh_{cg} can be calculated as

$$\Delta h_{cg} = (\pi/32)A$$

During one wave period the center of gravity drops by $2\Delta h_{cg}$ and energy is transformed from potential energy to kinetic energy:

$$\Delta E_{pot} = Mg2\Delta h_{cg} = MgA\frac{\pi}{16}$$

In our example this results in an energy $\Delta E_{pot} = 2 \times 10^8$ J or 78kW h or 10 kgCE of this wave front.

The power density of a wave per unit length of the wave front is then:

$$\frac{\mathrm{d}\bar{P}}{dx} = \frac{\Delta E_{pot}}{b} \cdot f_w = \frac{MgA\pi f_w}{16b} = \frac{\rho A^2 g \lambda_w f_w}{16}$$

$$\text{with } \lambda_w = \frac{g}{2\pi f_w^2} \quad \Rightarrow$$

$$\frac{\mathrm{d}\bar{P}}{dx} = \frac{1}{32\pi}\rho g^2 \frac{A^2}{f_w} \tag{4.1.5}$$

For an amplitude of $A = 1$ m and a wave frequency of $f_w = 0.5$ Hz the linear power density is about

$$\Rightarrow \mathrm{d}\bar{P}/\mathrm{d}x \approx 2000\,\mathrm{W\,m^{-1}}$$

For a 1 GW wave power station this would require a coastline length of $b =$ 500 km.

Figure 4.6 shows a map of the mean wave power per meter for the global oceans. As one can see, the energy density per unit length is especially low in coastal regions, where harvesting this energy would be more easy.

The first realized large-scale wave energy converter was Pelamis, a long, snake-like semi-submerged structure with a rated power of 750 kW. Its jointed, cylindrical sections move with the incoming waves and transfer this motion to hydraulic rams, charging hydraulic accumulators. Electricity is then generated by induction motors housed inside the structure [3]. A first power plant was installed off the coast of Portugal in 2008, but in the same year taken offline again. Most systems in use today are based on point absorber buoys. These buoys pick up the wave motion and transmit it to a shaft connected to a station in the sea bed, where its motion is converted to electricity. Small units are installed e.g. off the coast of Australia (CETO, rated power 240 kW per buoy) or of the United States (Azura Wave, rated power 20 kW per buoy). Another approach to convert wave energy to electricity is based on the oscillating water column. Here, the changing water level of incoming waves and tides pushes air through a wind turbine and, thus, produces electricity. However, so far no system has made it beyond a prototype phase.

4.2 Tidal Energy

Tides are created by asymmetries between gravitational and centrifugal acceleration of water on different sides of the Earth's surface relative to the Sun and the Moon.

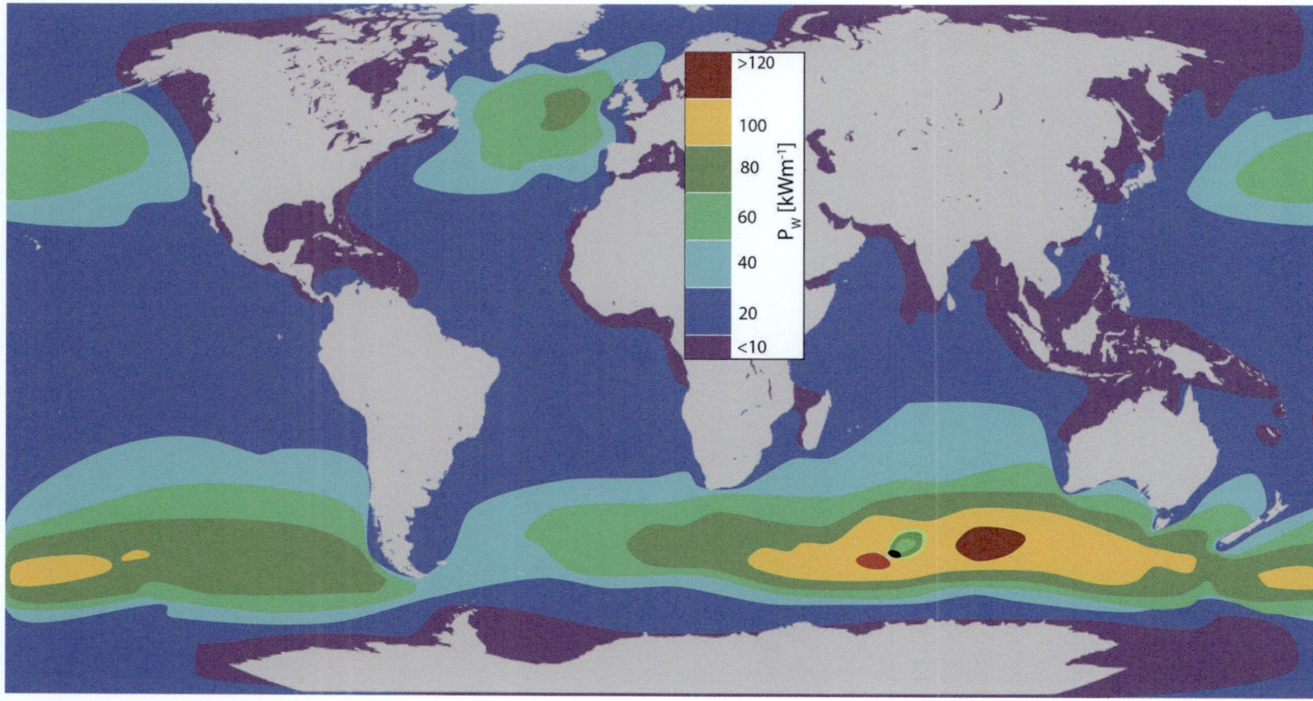

Fig. 4.6 Schematic global map of the average linear power density of ocean waves. More accurate maps can be found e.g. at Cornett [2]. The color scales from $>10\,\mathrm{kW\,m^{-1}}$ (purple) to $>120\,\mathrm{kW\,m^{-1}}$ (red)

Hence we can differentiate between solar and lunar tides. The complex overall tidal forces are a superposition of these two contributions depending on the relative constellation of Earth, Sun and Moon (e.g. spring tide).

4.2.1 Solar Tides

Solar tides result from the differences between the gravitational pull of the Sun and the centrifugal forces due to the orbit of the Earth around the Sun on the two sides of the Earth's surface closest to and furthest away from the Sun as shown in Fig. 4.7. On the side facing towards the Sun, the gravitational pull is larger than the centrifugal force, whereas on the opposite side the centrifugal force is larger than the gravitational pull. Only in the center of the earth both forces cancel each other exactly. There, the average centrifugal acceleration of the Earth on its orbit around the Sun is determined by the Sun-Earth distance ($r_{SE} = 1\text{AU} = 1.5 \times 10^{15}$ m) and the angular velocity ($\omega = (2\pi)/yr \approx 2 \times 10^{-7}\text{s}^{-1}$) of the Earth's yearly orbit:

$$\bar{a}_{cf} = r_{SE}\omega^2 = 6 \times 10^{-3}\,\text{m s}^{-2}$$

This average value varies throughout the Earth by

$$\Delta a_{cf} = 2R_{♁}\omega^2 = 2.5 \times 10^{-7}\,\text{m s}^{-2}$$

The average gravitational acceleration caused by the mass of the Sun on the other hand is

$$\bar{g}_{\odot} = \frac{GM_{\odot}}{r_{SE}^2} \qquad (= \bar{a}_{cf} \text{ only in the center of the Earth})$$

with a variation on both sides of the Earth of $g_{\odot}$:

$$|\Delta g_{\odot}| = \left|\frac{dg_{\odot}}{dr_{SE}}\right| \cdot 2R_{♁} = 2\frac{GM_{\odot}}{r_{SE}^3} \cdot 2R_{♁} = 1 \times 10^{-6}\,\text{m s}^{-2}$$

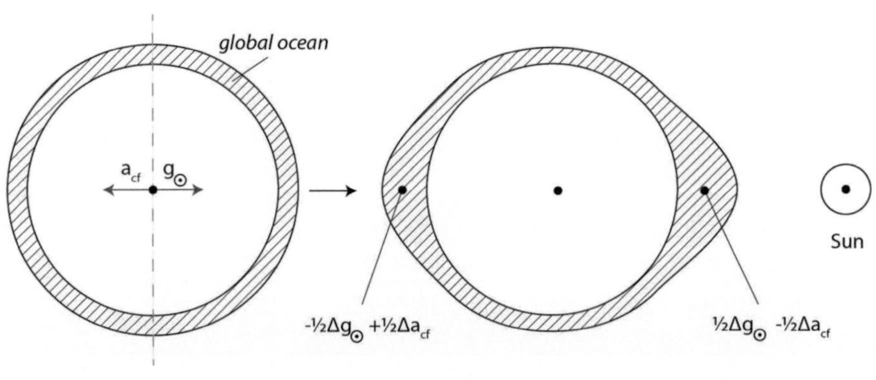

Fig. 4.7 Solar tides under the assumption of a global ocean expanding around the Earth

Since $\Delta a_{cf} \approx (1/4)\Delta g_{\odot}$, the gradient of the gravitational pull dominates the solar tides. Assuming a global ocean as the simplest approximation, this would give rise to two tidal waves on the opposite sides of the Earth with a height difference between low and high tide of only up to 20 cm. Due to the Earth's daily rotation we obtain a high tide period of 12 h at a given point on the Earth's surface.

4.2.2 Lunar Tides

The calculation of the lunar tides' height follows the same approach, with one important difference: Due to the large mass of the Sun and the large distance between Sun and Earth we could assume that the position of the center of mass of the Sun-Earth system coincides with the center of the Sun. For the Earth-Moon system, however, we have to take into account that its center of mass deviates significantly from the center of the Earth, as illustrated in Fig. 4.8.

The gradient of gravitational acceleration caused by the Moon is:

$$\Delta g_M = 2\frac{Gm_M}{r_{EM}^3} \cdot 2R_{♁} = 233 \times 10^{-6}\,\mathrm{m\,s^{-2}}$$

The corresponding calculation of the gradient of centrifugal force takes the position of the center of mass and the resulting additional centrifugal acceleration on the side facing away from the Moon (angular orbit frequency: $\omega = 2\pi/27.3\,\mathrm{d}$) into account:

$$\Delta a_{cf} \approx (R_{♁} + 4650\,\mathrm{km})\omega^2 = 8 \times 10^{-5}\,\mathrm{m\,s^{-2}}$$

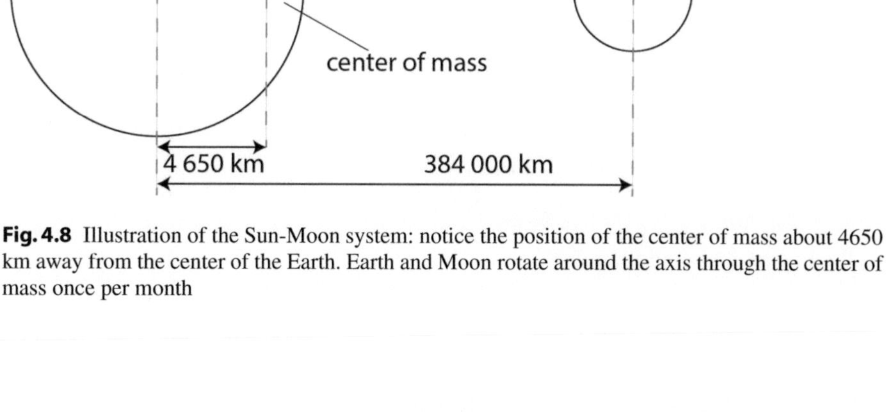

Fig. 4.8 Illustration of the Sun-Moon system: notice the position of the center of mass about 4650 km away from the center of the Earth. Earth and Moon rotate around the axis through the center of mass once per month

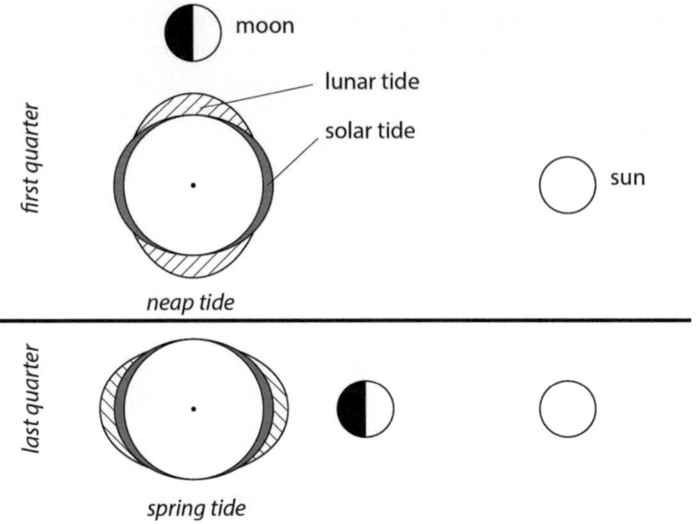

Fig. 4.9 Superposition of the solar and lunar tides for *neap* and *spring tides*

From these simple considerations we can see that the variation of forces through the Earth and therefore, also the lunar tides for a global ocean (50 cm) are larger than the solar tides (20 cm). Both contributions locally depend on the relative constellation of Sun, Earth, and Moon, as shown in Fig. 4.9, giving rise to periodic phenomena such as neap and spring tides.

The still rather small average amplitude of tidal waves for a global ocean can be locally enhanced when a tidal wave enters into a narrow, tapered channel, and also by local $\lambda/4$-*resonances*. As shown in Fig. 4.10, these occur when a tidal wave with a wavelength λ enters a coastal bay with a length of $\Delta x = \lambda/4$ perpendicular to the coast line. The distance $\lambda/4$ traveled by the incoming wave from the entrance to the end of the bay corresponds to a change of the wave phase by $\pi/2$. The wave is then reflected by the fixed end of the bay, giving rise to an additional phase change π. Finally, the reflected wave undergoes a second phase change of $\pi/2$ until it reaches again the entrance of the bay, where it interferes constructively with the next incoming wave:

$$\frac{\pi}{2} + \pi + \frac{\pi}{2} \Rightarrow 2\pi \text{ phase difference}$$

The Bay of Fundy in Canada shows one of the highest tides in the world. The bay has a length of L≈300 km and a depth of about d = 75 m. This corresponds to the limit of a shallow water wave for a tidal wave with a wavelength of about 1000 km. We can use this to estimate the time delay between the wave entering the bay and leaving it again:

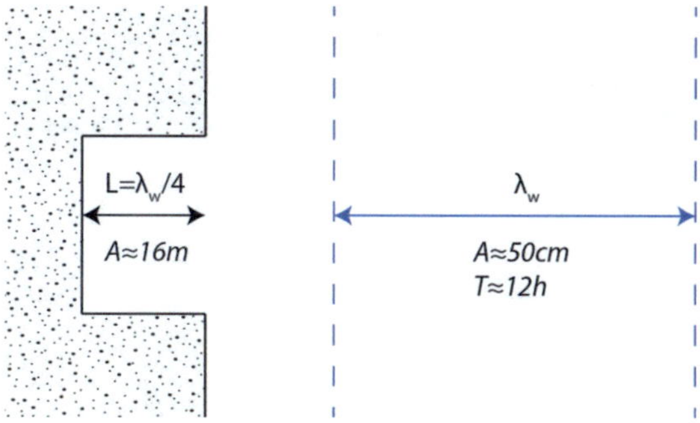

Fig. 4.10 Schematic illustration of the $\lambda/4$ resonance for a tidal wave with amplitude A and a period of $T = 12\,\text{h}$

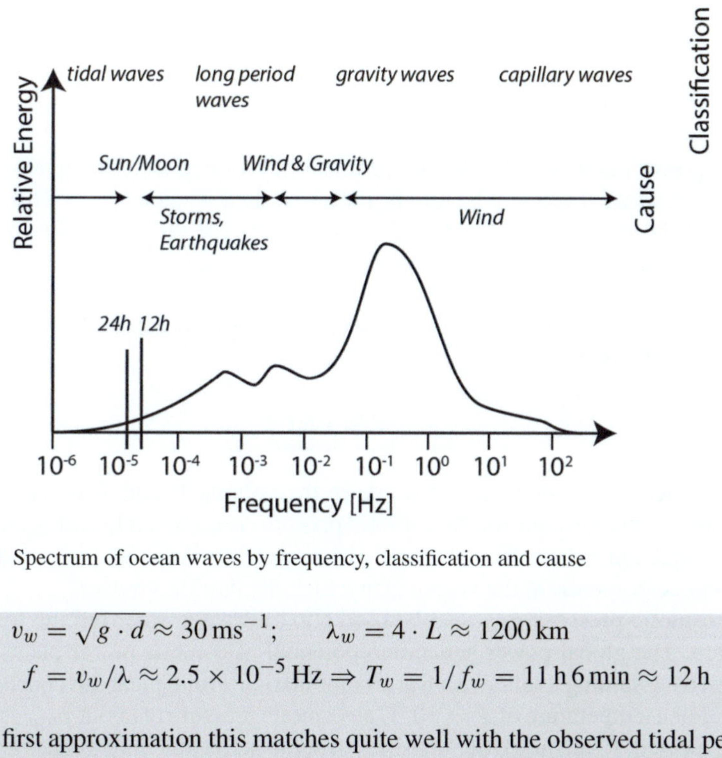

Fig. 4.11 Spectrum of ocean waves by frequency, classification and cause

$$v_w = \sqrt{g \cdot d} \approx 30\,\text{ms}^{-1}; \qquad \lambda_w = 4 \cdot L \approx 1200\,\text{km}$$

$$f = v_w/\lambda \approx 2.5 \times 10^{-5}\,\text{Hz} \Rightarrow T_w = 1/f_w = 11\,\text{h}\,6\,\text{min} \approx 12\,\text{h}$$

To a first approximation this matches quite well with the observed tidal period of around 12.4 h.

Figure 4.11 provides a qualitative summary of the relative energy provided by ocean waves of different origin as a function of the wave frequency. Tidal waves appear as delta functions with periods of 12 and 24 h at the low frequency end of the spectrum.

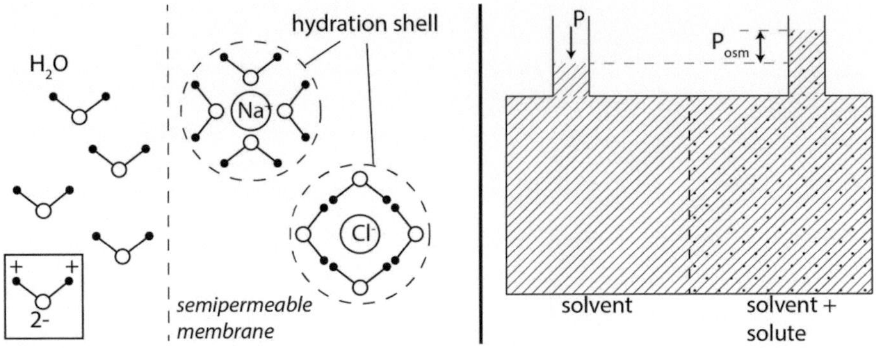

Fig. 4.12 Left: Polar water molecules decrease the energy of ions of the solute (NaCl) by forming a hydration shell around them. Right: Working principle of an osmosis power plant using the osmotic pressure: solvent molecules will diffuse into the second chamber with a higher solute concentration to further dilute the solute, giving rise to an osmotic pressure difference p_{osm}

4.3 Osmosis Power

Osmosis power plants use the difference in salinity of e.g. (salty) ocean water and (sweet) river water. Their key feature is a semi-permeable membrane, which is permeable for the solvent (i.e. H_2O) and in-permeable for the solute (i.e. NaCl). A simple osmotic cell consists of two compartments separated by such a membrane (cf. Fig. 4.12).

Across the membrane, an osmotic pressure develops, which can be estimated by the van-'t-Hoff equation:

$$\boxed{p_{osm} V = \nu R T}$$

Here, ν is the number of moles of solute in the volume V and R is the universal gas constant. This formula for the osmotic pressure is completely analogous to the ideal gas equation, where the solute ions correspond to the ideal gas particles and the solvent corresponds to the vacuum (in which the particles move).

This osmotic pressure p_{osm} can be used e.g. to drive a water turbine for power generation. The global power generation potential of osmotic power plants can be estimated by assuming a salt/sweet water combination with typical salt concentration of 3.5%. For a temperature of $T = 10\,°\mathrm{C}$ an osmotic pressure of about $p_{osm} \approx 20$ bar can be expected. Realistic power densities of semi-permeable membranes are about 3 W/m^2. Taking into account the limited sweet water resources (only 2.6–3.5% of the Earth's water is sweet water, with most of this in the form of snow and ice) we can assume a realistic osmosis power potential for Germany of about 40 MW and worldwide of about 60 GW.

The first osmosis power plant prototype worldwide was opened in 2010 close to Oslo in Norway and was designed for an electric power of 10 kW. It was decommissioned again in 2013.

Further Reading

- Drew, B., Plummer, A.R., Sahinkaya, M.N.: A review of wave energy converter technology. J. Pow. Eng., Part A, **223,8**, 887–902 (2009).
- López, I., Andreu, J., Ceballos, S., Martínez de Alegría, I., Kortabarria, I.: Review of wave energy technologies and the necessary power-equipment. Renew. Sustain. Energy Rev. **27**, 413–434 (2013).

Some examples of wave energy converters:

- CETO: carnegiece.com/ceto-technology
- Azura wave: http://azurawave.com
- Statkraft Osmotic: www.power-technology.com/projects/statkraft-osmotic/.

References

1. Elmore, W.C., Heald, M.A., Stumpf, F.B.: Physics of waves. Jour. Acoust. Soc. Am. **81**, 204–204 (1987)
2. Cornett, A.: A global wave energy resource assessment. Sea Technol. **50**, 59–64 (2009)
3. Yemm, R., Pizer, D., Retzler, C., Henderson, R.: Pelamis: experience from concept to connection. Phil. Trans. Roy. Soc. A **370**, 365–380 (2012)

5 Wind Energy

Abstract

In this chapter the principles of wind, its energy content and the forms of conversion are explained. The maximum efficiency of turbines is derived, followed by a discussion of different types of real wind engines, their loss mechanisms and possible routes for optimization.

5.1 General Considerations

For the existence of weather phenomena such as wind a planetary atmosphere is necessary. In our solar system, not all planets have a sufficiently thick atmosphere or do not even have one at all. Mercury for example has an "atmosphere" thinner than today's best achievable vacuum. It seems that all particles that could build up an atmosphere are blown away by the solar wind. Jupiter on the other hand as a gaseous planet consists nearly only of one thick atmosphere.[1] The different colorful cloud bands and the distinctive great red spot show the vivid motion in this atmosphere.

In general wind is a directed motion of atmospheric masses. On our Earth the wind pattern is mainly determined by three influences:

- Convection due to different surface temperatures
- Coriolis force due to Earth rotation ($\mathbf{F} = 2m\left(\mathbf{v}' \times \boldsymbol{\omega}\right)$, see below)
- Local high and low pressure zones

Thus, wind energy is just a transformation of solar energy, caused by the solar irradiation and the inhomogeneous temperature of the Earth's surface. The global energy in wind corresponds to about 2% of the solar irradiation energy. Apart from local and temporal aberrations we can simplify the system to a more or less steady,

[1] Of course this is an exaggeration: at the higher pressures inside the gases become liquid. Whether Jupiter has a solid core or not, however, has not been proven yet.

M. Stutzmann and C. Csoklich, *The Physics of Renewable Energy*, Graduate Texts in Physics,
https://doi.org/10.1007/978-3-031-17724-8_5

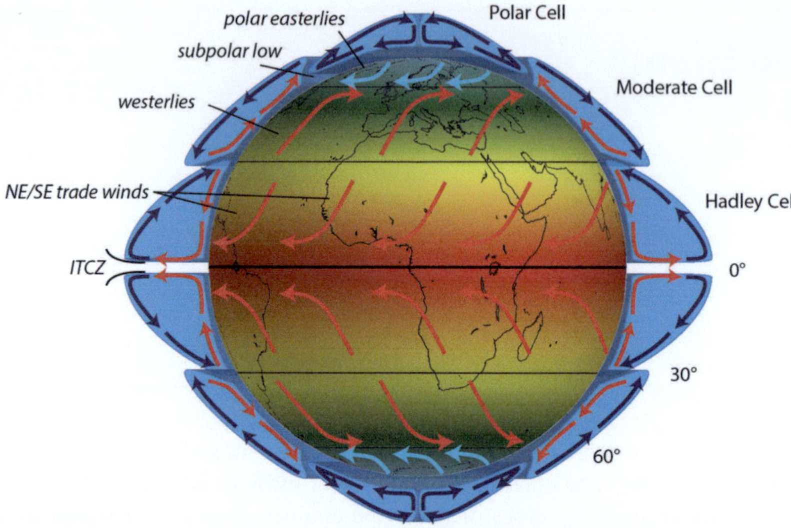

Fig. 5.1 Schematic view of the global wind patterns on Earth. See text for details

global wind pattern as shown in Fig. 5.1. The high solar irradiation at the equator heats the air there, so that it rises upwards and flows towards the poles in six, large convection rolls. The air masses coming from the equator start sinking already at about 30° N/S. Cool air, coming from the poles and flowing towards the equator on the Earth surface warm up on their way and rise at about 60° N/S.

The air masses moving in these convection rolls develop an increasing component of their surface velocity $\mathbf{v}'$ perpendicular to the rotational axis of the Earth the closer they are to the poles. Thus, they are increasingly influenced by the Coriolis force ($\mathbf{F}_C$, cf. Fig. 5.2). On the northern hemisphere, poleward winds are deflected to the right, on the southern hemisphere to the left. At the equator, the influence of $\mathbf{F}_C$ is only relevant for rising air. Altogether, this leads to an east/west deflection of the convection pattern. Note that for a given convection cell the direction of deflection due to the Coriolis force is opposite for the air high up in the atmosphere (leading to the formation of jet streams) and the air directly above the Earth surface which we realize as global winds.

Apart from the rather small influence of local high and low pressure zones, which are more important for the (short time) local weather, we can summarize these effects with the development of three cells[2]:

Tropical or Hadley Cell

The deflected air masses cause rather steady NE and SE winds[3] called *trade winds* or *Hadley circulation*. Since these two wind belts meet close to the equator, there

[2] The free, educative website earth.nullschool.net displays the life movement of Earth's wind and ocean streams.

[3] The direction of winds is named after the point of compass from where they come.

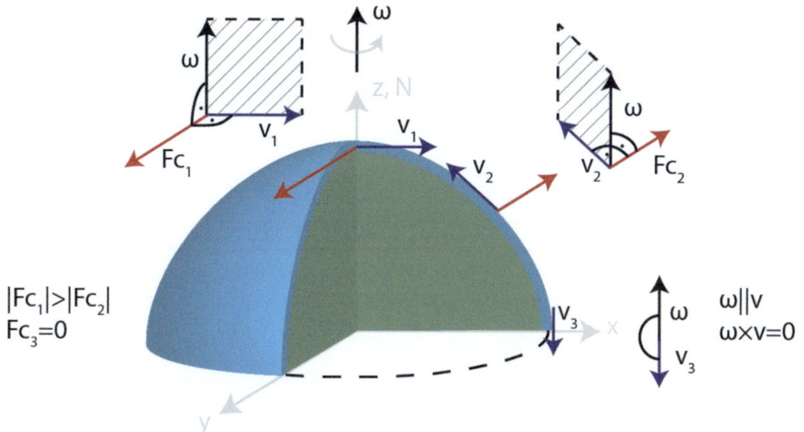

Fig. 5.2 Illustration of the Coriolis force at different positions on the Earth surface

exists a global equatorial low pressure channel, called the *intertropical conversion zone (ITCZ)*, which varies in its exact position during the year.

Moderate or Ferrel Cell
Here the Coriolis deflection causes a west wind belt, the *westerlies*. Especially in higher altitudes, close to the tropopause, there exist very fast, narrow and meandering west wind belts, called the *jet-stream*.

Polar Cell
Here the cold air from the poles flows towards the equator and warms up, until at around 60° N/S it rises again and moves backwards. The so called *Polar front* between Polar and Ferrel cell is very unstable due to turbulences and leads to a meandering border line in 4–6 so called *Rossby waves*.

5.2 Energy Content of Wind

After discussing the global wind patterns of the Earth, we now take a look at the energy stored in wind. Since wind is a motion of particles, we consider the kinetic energy of streaming air of a cylindrical volume element dV (cf. Fig. 5.3):

$$dE_{kin} = \frac{1}{2} dm v^2$$

We can rewrite this expression by using the following relations for the mass element:

$$dm = \rho dV \qquad dV = A dx \qquad v = \frac{dx}{dt}$$

$$\Rightarrow dm = \rho A v \, dt$$

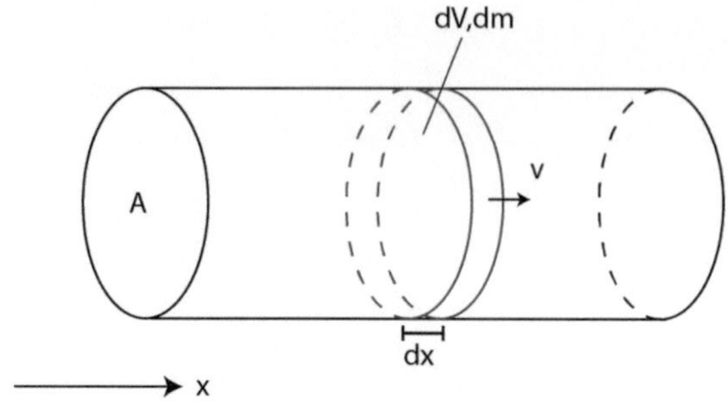

Fig. 5.3 Geometry for calculating the energy content of wind: a volume element $\mathrm{d}V = A\mathrm{d}x$ containing the mass $\mathrm{d}m$ is moving with wind velocity v along a cylindrical cross section with area A

The mass flux $\dot{m}$ is thereby defined as:

$$\frac{\mathrm{d}m}{\mathrm{d}t} = \dot{m} = \rho\, A\, v$$

This leads to the final relations for kinetic energy and the power density per unit area of streaming air:

$$\mathrm{d}E_{kin} = \frac{1}{2}\rho v^3\, A\, \mathrm{d}t$$

$$\boxed{\frac{P}{A} = \frac{1}{A}\frac{\mathrm{d}E_{kin}}{\mathrm{d}t} = \frac{1}{2}\rho v^3} \tag{5.2.1}$$

The density of air at normal pressure (i.e. $p = 1$ bar) at sea level is $\rho \approx 1.2\,\mathrm{kgm^{-3}}$. It can be calculated from the ideal gas equation:

$$\rho = \frac{p \cdot M}{R \cdot T} = \frac{p}{R_S \cdot T}$$

where R_s is the specific gas constant R/M and T the absolute temperature.

- Temperature: For a temperature variation from $-25\,°\mathrm{C}$ to $+35\,°\mathrm{C}$[4] the density varies only by $\Delta\rho = 0.27\,\mathrm{kg\,m^{-3}}$ from 1.42 to $1.15\,\mathrm{kg\,m^{-3}}$
- Pressure: The air pressure varies mostly with height, as can be derived from the barometric height formula. It ranges from normal pressure at sea level 1013.25 hPa to about 325.4 hPa at the top of Mt. Everest.

However, in practice the contributions by the variation of ρ are negligible, since a wind turbine is most likely installed at a fixed height. Therefore, the arriving wind power density is mainly affected by variations of the velocity. From these simple

[4] At normal pressure at sea level.

considerations we can see that the energy content of wind is determined by the long-term average wind velocity **v** and is proportional to v^3. Thus, wind energy harvesting systems should be located in areas with a high average wind speed over the year.

For a wind speed of $v = 10\,\mathrm{m\,s^{-1}}$ and a regular air density of $\rho = 1.2\,\mathrm{kg\,m^{-3}}$ we find a power density of $P/A = 600\,\mathrm{W\,m^{-2}}$. A small wind turbine with a rotor area of $A = 100\,\mathrm{m^2}$ receives an incoming power of $P = 60\,\mathrm{kW}$.

The distribution of wind speed v varies strongly as a function of time and place (cf. Fig. 5.4), whereas its variation of v with height is weaker and dependent on the average wind friction with the surface. An empirical relation for this dependence is:

$$v = v_{10m}\left(\frac{h}{10m}\right)^{f^*}$$

where f^* is a surface-dependent friction coefficient: it ranges from $f^* \approx 0.12$ for flat surfaces (e.g. the sea) to $f^* \approx 0.4$ in cities with high buildings. The height h above the surface is normalized to 10 m. Typical wind velocities in Europe lie between $0 < v < 30\,\mathrm{m\,s^{-1}}$. In storms, speeds up to $60\,\mathrm{m\,s^{-1}}$ (i.e. $200\,\mathrm{km\,h^{-1}}$) can be reached. However, the commercial use of wind energy begins for a minimum wind speed of $4\,\mathrm{m\,s^{-1}}$. As shown in Fig. 5.5, the relative wind speed distribution at a given site can be approximated quite well by the so-called *Weibull* or *Rayleigh distributions*:

$$f_{Weibull}(v) = \frac{k}{a}\left(\frac{v}{a}\right)^{k-1}\exp\left[-\left(\frac{v}{a}\right)^k\right] \tag{5.2.2}$$

$$f_{Rayleigh}(v) = \frac{\pi}{2}\frac{v^2}{\bar{v}^2}\exp\left(-\frac{\pi}{4}\frac{v^2}{\bar{v}^2}\right) \tag{5.2.3}$$

5.3 Efficiency of Wind Turbines

Wind turbines are the most important technology to convert wind energy into electricity. Though many different designs and realizations can be found, the ideal efficiency of a wind engine can be calculated independent of the type of turbine, shown first

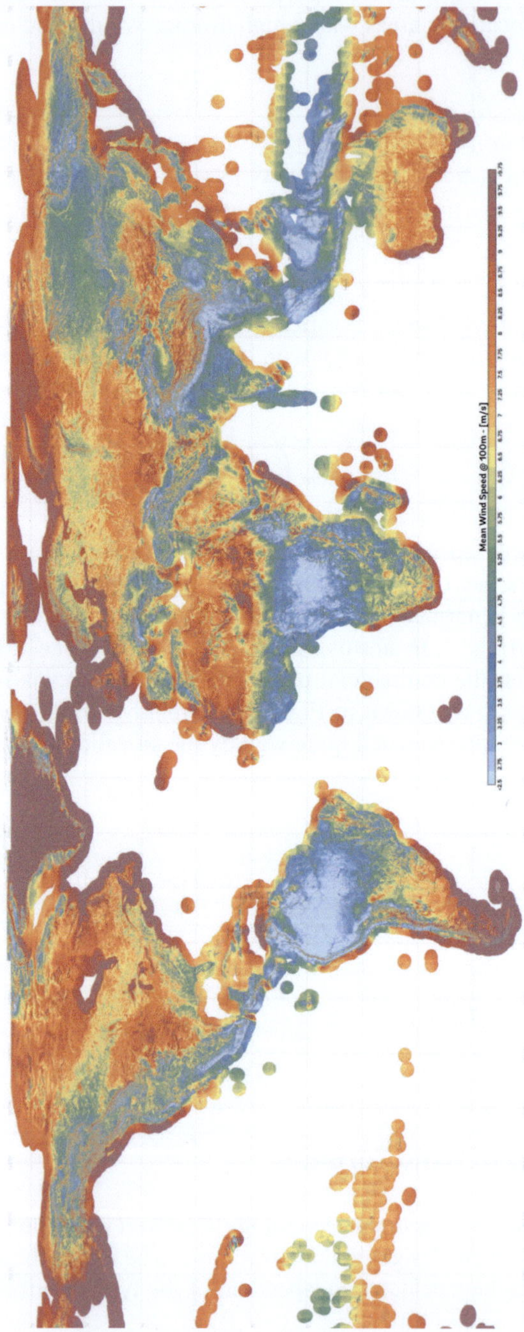

Fig. 5.4 Wind resource map showing the mean wind speed at 100 m above surface. Map obtained from the Global Wind Atlas 3.1, a free, web-based application developed, owned and operated by the Technical University of Denmark (DTU) [1]. The permission to reprint is greatfully acknowledged

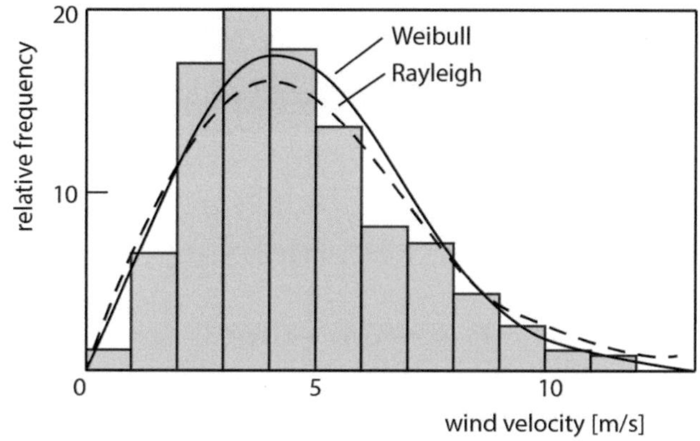

Fig. 5.5 A typical local wind speed distribution with different mathematical approximations as described in the text

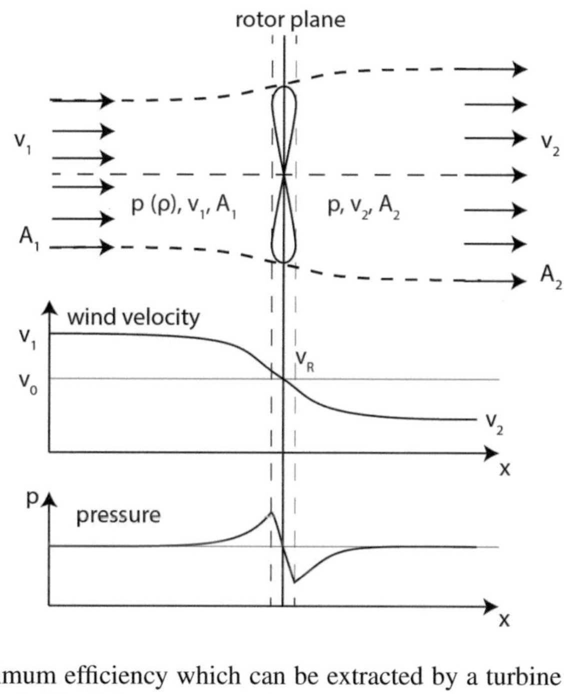

Fig. 5.6 Generalized ansatz for the efficiency calculation for wind engines according to Betz. A_1 and v_1 are the area and wind velocity in front of the rotor plane, A_2 and v_2 behind the rotor plane. The pressure p and therefore also the density ρ remain unchanged sufficiently far from the rotor

by Betz [2]. It limits the maximum efficiency which can be extracted by a turbine based on the parameters shown in Fig. 5.6.

Let a wind engine interact with incoming air of speed v_1 and density ρ_1 in an area A_1. The pressure in front of the wind turbine and in some distance behind it will be the same. This means, that also the density of air is constant. Thereby the speed has to drop after passing the turbine to $v_2 < v_1$ and flows through a larger area $A_2 > A_1$.

A given mass element $dm_1 = \rho dV = \rho A_1 dx = \rho A_1 v_1 dt$ is conserved when going through the rotor plane. Assuming that $\rho_1 = \rho_2 = \rho$ this gives (continuity equation):

$$\rho A_1 v_1 \mathrm{d}t = \rho A_2 v_2 \mathrm{d}t \Rightarrow \boxed{v_1 A_1 = v_2 A_2}$$

The wind velocity v_R in the rotor plane can be calculated using the Bernoulli equation:

$$v_R = \frac{1}{2}(v_1 + v_2)$$

According to Sect. 5.1 the incoming wind power is:

$$P_1 = \frac{P}{A} A_1 = \frac{1}{2} \rho A_1 v_1^3$$

Accordingly the outgoing wind power is:

$$P_2 = \frac{P}{A} A_2 = \frac{1}{2} \rho A_2 v_2^3$$

The power extracted by the wind engine is then:

$$P_R = P_1 - P_2 = \frac{1}{2}\rho(A_1 v_1^3 - A_2 v_2^3) = \frac{1}{2}\rho(\underbrace{A_1 v_1}_{=X} v_1^2 - \underbrace{A_2 v_2}_{=X} v_2^2) = \frac{1}{2}\rho A_1 v_1 (v_1^2 - v_2^2)$$

Alternatively, using $\rho A v = \mathrm{d}m/\mathrm{d}t$ this can be written as:

$$\boxed{P_R = \frac{\dot{m}}{2}\left(v_1^2 - v_2^2\right)} \tag{5.3.1}$$

Thus, P_R is proportional to the air mass flow and the change in kinetic energy ($\propto v^2$). In the rotor plane we have (using $A_R = A_1$):

$$v = v_R = \frac{1}{2}(v_1 + v_2) \qquad \dot{m} = \rho A_R v_R$$

$$\Rightarrow \dot{m} = \frac{1}{2}\rho A_1 (v_1 + v_2)$$

$$P_R = \frac{1}{4}\rho A_1 (v_1 + v_2)(v_1^2 - v_2^2) = \frac{1}{4}\rho A_1 v_1^3 \left(1 + \frac{v_2}{v_1}\right) \cdot \left(1 - \left(\frac{v_2}{v_1}\right)^2\right)$$

Since the incoming power is $P_1 = \frac{1}{2}\rho A_1 v_1^3$ the **efficiency c_P** of the (ideal) wind engine is (Fig. 5.7):

$$\boxed{c_P := \frac{P_R}{P_1} = \frac{1}{2}\left(1 + \frac{v_2}{v_1}\right) \cdot \left[1 - \left(\frac{v_2}{v_1}\right)^2\right]}$$

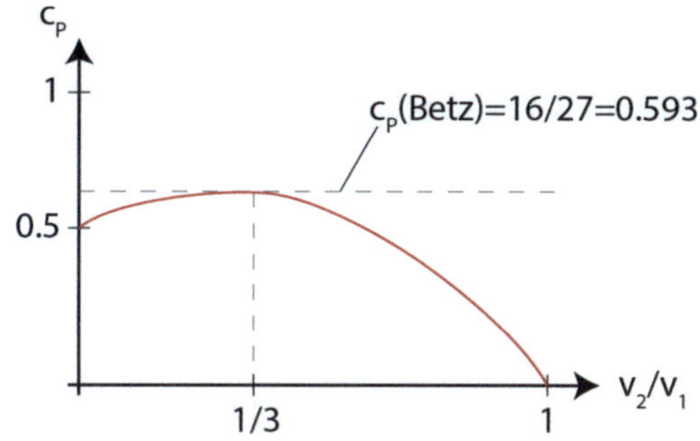

Fig. 5.7 Betz efficiency of a generalized wind engine versus the speed ratio v_2/v_1. The maximum of the efficiency is about 60%, occurring for a speed ratio of 1/3

Derivation of the condition $v_R = \frac{1}{2}(v_1 + v_2)$ for the wind velocity v_R in the rotor plane using the Bernoulli equation $p + \frac{1}{2}\rho v^2 = const.$ (Indices: 1: far in front of the rotor, 2, far behind it, $-R$: just before, $+R$: right after the rotor).

$$p_1 + \frac{1}{2}\rho v_1^2 = p_{-R} + \frac{1}{2} + \rho v_{-R}^2 \tag{5.3.2}$$

$$p_{+R} + \frac{1}{2}\rho v_{+R}^2 = p_2 + \frac{1}{2} + \rho v_2^2 \tag{5.3.3}$$

Here $v_{-R} = v_R = v_{+R}$ can be assumed in good approximation. Subtraction of these two equations yields with $p_1 = p_2$

$$\frac{1}{2}\rho\left(v_1^2 - v_2^2\right) = p_{-R} - p_{+R} \tag{5.3.4}$$

The force acting on the rotor due to the pressure difference is:

$$F = A_R(p_{-R} - p_{+R})$$

On the other hand the force is given by the time derivative of the momentum:

$$F = \frac{\mathrm{d}m}{\mathrm{d}t}(v_1 - v_2) = A_R\rho v_R(v_1 - v_2)$$

This results in:

$$\Rightarrow p_{-R} - p_{+R} = \rho v_R(v_1 - v_2)$$

Inserting this into (5.3.4) above finally gives:

$$v_R = \frac{1}{2}(v_1 + v_2) \tag{5.3.5}$$

5.4 Types of Rotors

5.4.1 Drag-Type Rotors

Drag-type rotors use the drag force of solid bodies in streaming air with velocity v:

$$F_W = c_W \frac{1}{2} \rho \, v^2 A$$

where A is the (effective) area perpendicular to the wind direction. The drag coefficient c_W depends on the shape of the body as shown in Table 5.1.

A simple drag rotor is sketched in Fig. 5.8. The maximum drag force on the upper panel is:

$$F_W = c_W \frac{1}{2} \rho A (v_1 - u)^2$$

Table 5.1 Drag coefficients in air for bodies with different shapes

Shape	c_W	
Stream-lined	0.03	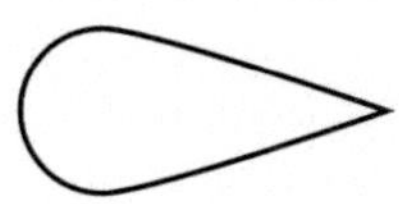
Wing-shaped	0.03-0.1	
Sphere	0.4	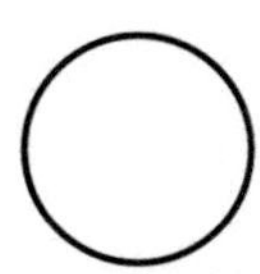
Semi-sphere	0.8	
Plane	1.2	
Hollow semi-sphere	1.4	

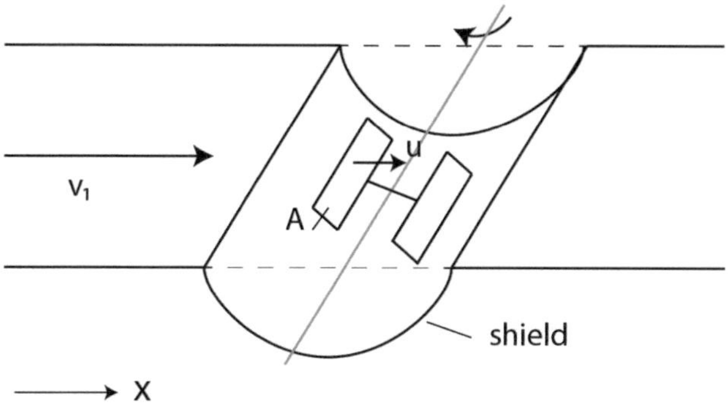

Fig. 5.8 Simple example of a drag-type wind engine. Two panels with area A can rotate with a circumferential velocity u around a common axis perpendicular to the direction of the wind velocity v_1. A trench acting as wind shield for the lower half of the engine breaks the symmetry of the panel arrangement

where u is the circumferential moving speed of the panel. This gives a power taken from the wind by the rotor of:

$$P_R = \frac{\mathrm{d}E}{\mathrm{d}t} = \frac{\mathrm{d}}{\mathrm{d}t} F_W \mathrm{d}x = F_W \cdot u$$

The efficiency of this drag-drag type wind engine is accordingly:

$$c_p = \frac{P_R}{P_1} = \frac{c_W \frac{1}{2}\rho A (v_1 - u)^2 \cdot u}{\frac{1}{2}\rho A \cdot v_1^3}$$

$$\boxed{c_P = c_W \left(1 - \frac{u}{v_1}\right)^2 \cdot \left(\frac{u}{v_1}\right)} \tag{5.4.1}$$

The maximum of c_P is again obtained for $u/v_1 = 1/3$. The maximum value is $c_{P,max} = c_W(4/27)$, so that for $c_W \leq 2$ the maximum efficiency of drag type wind engines is $\leq 30\%$. This is a factor of two lower than the ideal Betz efficiency, showing that drag-type rotors are sub-optimal for wind energy harvesting.

5.4.2 Lift-Type Rotors

Better efficiencies, close to the Betz limit, can be obtained for lift-type rotors. These use the lift force of a wing-shaped rotor:

Due to the specific wing profile, the air above the wing has to pass a longer trajectory than the air below the wing. As shown in Fig. 5.9, this results in a larger

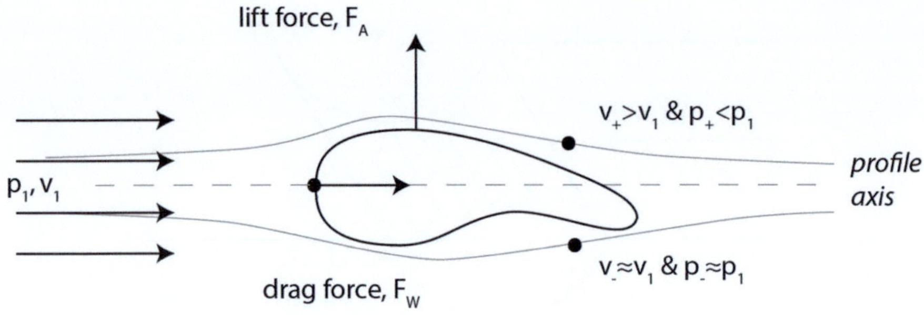

Fig. 5.9 Schematic drawing of the air flow around a lift-type rotor and the resulting difference in pressure above and below the wing

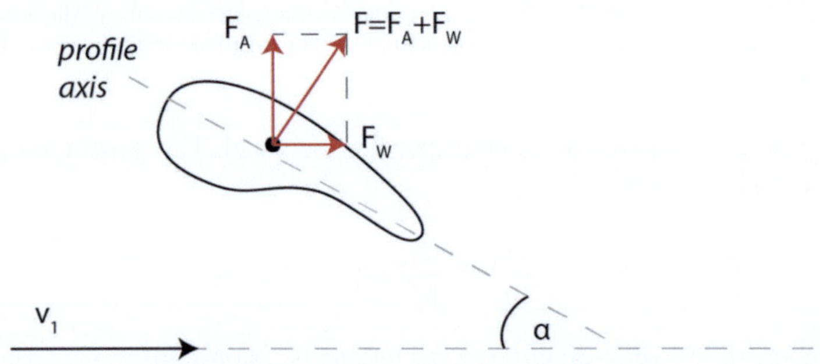

Fig. 5.10 Forces acting on a wing with a finite angle of attack α relative to direction of wind

air velocity $v_+ > v_1$ above the wing and thus, according to the Bernoulli equation, in a lower pressure $p_+ < p_1$ there. This gives rise to a lift force F_A according to the following equation, where $A_\parallel$ is the area of the wing parallel to the wind and c_A the so-called *lift coefficient*, defined in a similar way as the drag coefficient above:

$$F_A = (p_1 - p_+)A_\parallel \tag{5.4.2}$$

$$= c_A \frac{1}{2} \rho A_\parallel v_1^2 \tag{5.4.3}$$

An important parameter for the operation of lift-type rotors is the *angle of attack* α between the wind direction and the *profile axis* of the wing (cf. Fig. 5.10).

A variation of α allows to change the effective lift coefficient c_A, which is accompanied by a smaller change of the drag coefficient c_W. For a given wing profile, this variation is documented in the so-called *polar profile plot*. A typical example of such a plot is shown in Fig. 5.11.

For a given angle of attack, c_A and thus the lift force vanish (stall condition). A second stall condition may occur for large angles of attack in very strong winds, if the angle of attack is not actively corrected. Then, with increasing wind velocity v_1 and for constant circumferential turning velocity u, the angle of attack will also

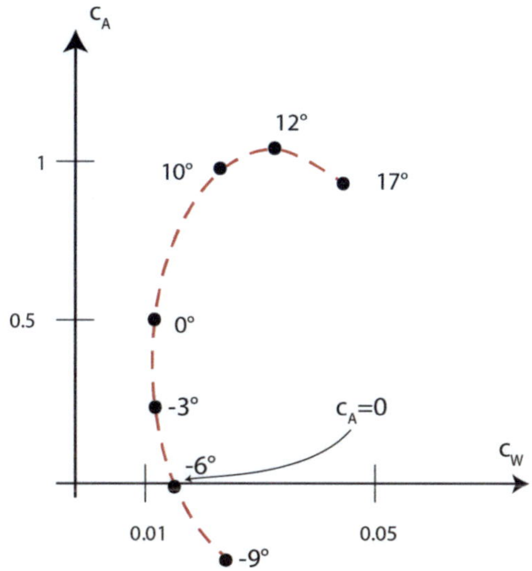

Fig. 5.11 Polar plot for an optimized profile, showing the dependence of c_A and c_W on the angle of attack. In particular, c_A vanishes at the stall angle ($-6°$ in this case)

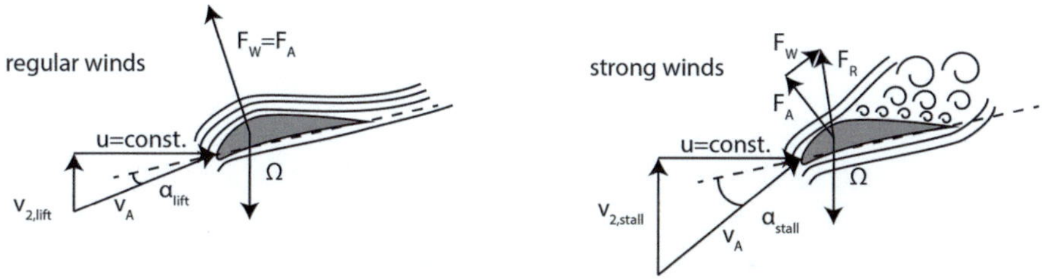

Fig. 5.12 Stall condition in strong winds, caused by a too large angle of attack

increase, as is shown in Fig. 5.12. This eventually causes the drag coefficient to strongly increase and the lift coefficient to decrease. As will be discussed further below, for lift-type rotors the wing or blade profile needs to vary along the blade to account for the different air velocities along the blade, i.e. from the center of the rotor to the tip of the blade. Also, in modern wind engines the angle of attack is continuously adjusted to optimize the lift coefficient and, thus, the lift force for different wind velocities. This is called *pitching*.

Another important parameter of a wing profile is the ratio of F_A and F_W, which is called *glide number* G[5]:

$$G := \frac{F_A}{F_W}$$

[5] Optimized profiles can reach glide numbers close to 100.

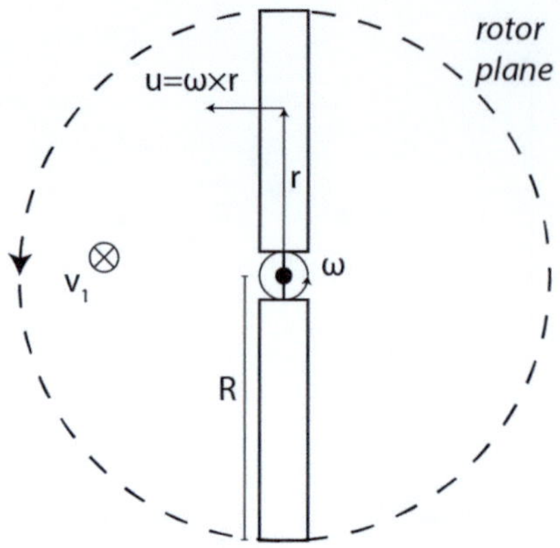

Fig. 5.13 Schematic drawing of a two-bladed lift-type rotor as seen from the direction of the incoming wind

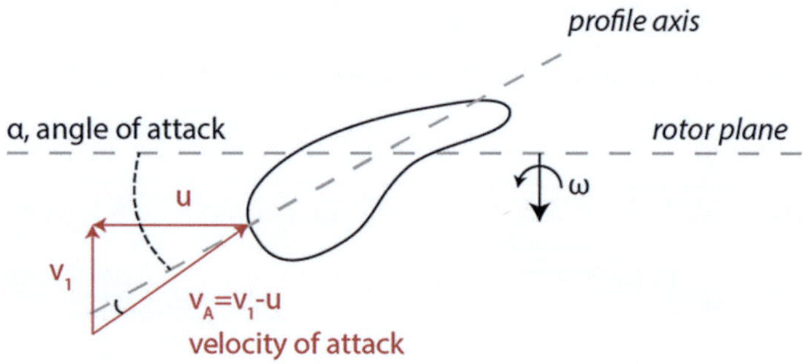

Fig. 5.14 View (rotated by 90°) from the top of the rotor plane showing the velocity and the angle of attack

For further, more quantitative aspects we have to take a closer look at the geometry of lift-type rotors. As shown in Fig. 5.13 for a two-blade rotor, the direction of the incoming wind (v_1) is perpendicular to the rotor plane. The same holds for the rotor's angular velocity ω. At a distance r from the rotor center, a blade section moves with the local circumferential velocity $u = \omega \cdot r$ within the rotor plane.

Turning the rotor plane by 90° and looking from the top (cf. Fig. 5.14), we see that the *velocity of attack* v_A, i.e. the velocity with which the air hits the blade front, is the vector sum of the wind velocity and the blade velocity, $\mathbf{v_A} = \mathbf{v_1} + \mathbf{u}$. The corresponding angle of attack is the angle between the profile axis and the direction of v_A.

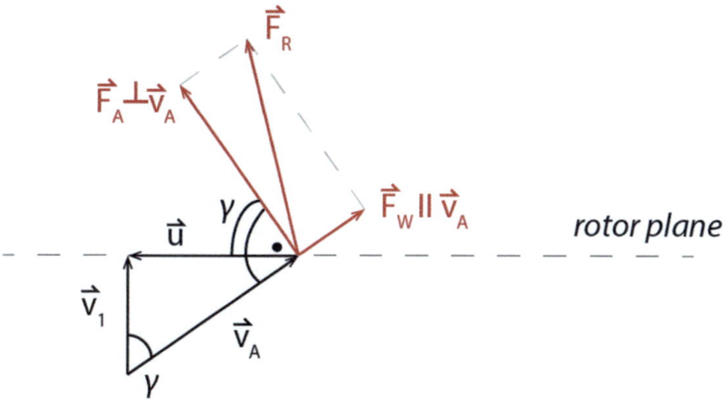

Fig. 5.15 Geometry of the lift force $\mathbf{F}_A$ and the drag force $\mathbf{F}_W$ perpendicular and parallel to the direction of the velocity of attack

The ratio

$$\boxed{\lambda := \frac{u}{v_1}} \tag{5.4.4}$$

is called *speed ratio*.

- For drag type rotors $\mathbf{v}_1$ and $\mathbf{u}$ are co-linear, so that $u \leq v_1 \Rightarrow \lambda \leq 1$. Optimized efficiency is reached for $\lambda \approx 1/3$
- For lift-type rotors $u(r) = \omega r$ can become much larger than $v_1 \Rightarrow \lambda > 1$. Depending on the *tip speed ratio* $\lambda_s = \frac{\omega R}{v_1}$ for the tip of the rotor $(r = R)$ one distinguishes:
  - $\lambda_S < 3$: slow rotors
  - $\lambda_S \geq 3$: fast rotors

The direction of the velocity of attack also defines the direction of the lift force $\mathbf{F}_A$ and the drag force $\mathbf{F}_W$, according to Fig. 5.15.

The total force on the rotor is then $\mathbf{F}_R = \mathbf{F}_A + \mathbf{F}_W \approx \mathbf{F}_A$ to a good approximation, because for good profiles the glide number $G = F_A/F_W >> 1$. Writing $\mathbf{F}_R = \mathbf{F}_{R,\parallel} + \mathbf{F}_{R,\perp}$ ($\parallel$ and $\perp$ to the rotor plane) we get

$$F_{R,\parallel} \approx F_A \cos(\gamma); \qquad F_{R,\perp} \approx F_A \sin(\gamma); \qquad \cos(\gamma) = \frac{v_1}{v_A}$$

The power harvested by a blade segment dr at a distance r from the rotor axis is then:

$$\mathrm{d}P(r) = \mathbf{dF}_{R,\parallel} \cdot u(r) = \omega r \cos(\gamma)\mathrm{d}F_A = \omega r \frac{v_1}{v_A}\mathrm{d}F_A$$

with

$$F_A = c_A \frac{1}{2}\rho A v_A^2 \qquad \Rightarrow \qquad \mathrm{d}F_A = c_A \frac{1}{2}\rho v_A^2 \underbrace{\mathrm{d}A}_{=b\cdot \mathrm{d}r}$$

where b is the breadth of the blade at position r. Together this gives:

$$\begin{aligned}\mathrm{d}P(r) &= \omega r \frac{v_1}{v_A} c_A \frac{1}{2}\rho v_A^2 b\, \mathrm{d}r \qquad |v_A^2 = v_1^2 + u^2 \\ &= c_A \frac{1}{2}\rho\omega v_1^2 \sqrt{1 + \frac{u^2}{v_1^2}}\, b\, r\, \mathrm{d}r \qquad |\lambda(r) = \frac{u(r)}{v_1}\end{aligned}$$

$$\Rightarrow \boxed{\mathrm{d}P(r) = \frac{1}{2} c_A \rho\omega v_1^2 \sqrt{1 + \lambda^2(r)}\, b \cdot r \cdot \mathrm{d}r} \qquad (5.4.5)$$

Then, a simple rotor with z blades of length R and constant breadth b with a varying lift coefficient $c_A(r)$ along the blade harvests a total wind power of:

$$P = \int \mathrm{d}P(r) = \frac{1}{2}\rho\omega v_1^2 b z \int_0^R \sqrt{1 + \lambda^2(r)}\, c_A(r) r\, \mathrm{d}r$$

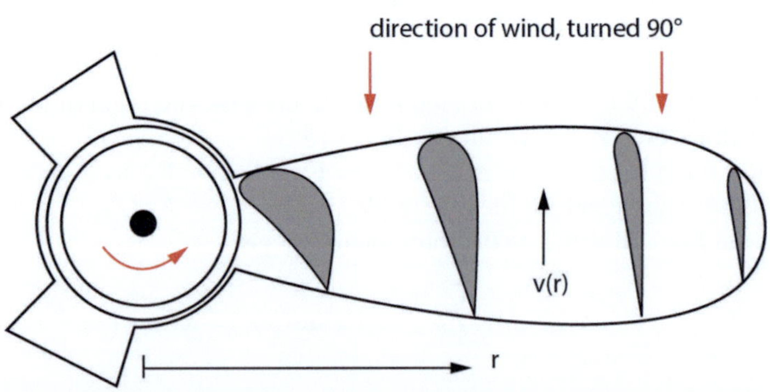

Fig. 5.16 Typical variation of the profile of a rotor blade between center and tip. The blade segments have been turned by 90° out of the rotor plane

- P is proportional to v_1^2 and ω
- $u = \omega r$ varies strongly from the beginning to the tip of a blade, so that the direction and size of the velocity of attack $\mathbf{v}_A$ change accordingly. Therefore, as shown in Fig. 5.16, the blade profile has to be adjusted along r and as a consequence, the lift coefficient c_A becomes a function of r.
- As already discussed in connection with the profile polar plot, the angle of attack changes c_A and c_W which can be used to optimize P for different v_1 ("pitching"). In particular c_W can be enhanced so much, that the rotor type changes from lift to drag.

To estimate the optimum number z of blades for a wind engine, consider a trapezoidal blade with an average breadth $\bar{b}$ at a distance $\bar{r} = 2/3\,R$. Then, the total *area coverage* of a rotor with z such blades is:

$$B = \frac{z\bar{b}}{2\pi\bar{r}}$$

The optimal coverage B_{max} depends on the *tip speed ratio* $\lambda_S := (\omega R)/v_1$: for the maximum rotor efficiency $c_P = 16/27$ one can then deduce:

$$B_{max} = \frac{8}{9}\frac{1}{c_A}\left[\left(\frac{\bar{r}}{R}\right)^2 \sqrt{1 + \frac{4}{9}\lambda_S^2 \left(\frac{\bar{r}}{R}\right)^2}\,\right]^{-1} \cdot \frac{1}{\lambda_S^2} \tag{5.4.6}$$

Thus, the optimum number of blades z has to be reduced with increasing tip speed ratio λ_S, which in turn needs to be adjusted to remain close to the Betz efficiency maximum.

5.4.3 New Types of Wind Engines

A very recent development and still in pre-industrial phase are bladeless wind engines based on oscillating towers (vortexbladeless.com) or energy kites, flying several hundreds of meters above ground. The interested readers are referred to the literature and provided links.

5.5 Optimization of Wind Turbines

In deriving the maximum Betz efficiency of wind engines, so far all losses have been neglected. In reality, there are three main loss mechanisms which have to be considered:

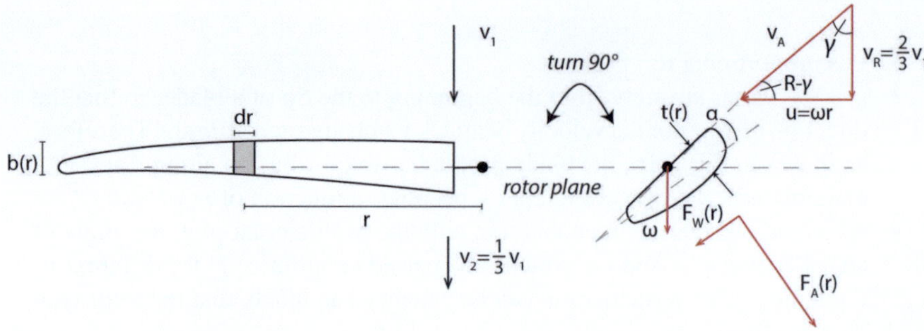

Fig. 5.17 Left: Blade breadth $b(r)$ and differential segment dr as a function of distance r from the center of the rotor. Right: Velocity triangle defined by the turning velocity u, the wind velocity v_R in the rotor plane, and the resulting velocity of attack v_A. v_R is shown for the optimum Betz condition $v_2 = v_1/3$. α is the angle of attack, $t(r)$ the effective blade thickness determining the friction force $F_W(r)$

profile losses: due to the non-zero c_W caused by friction.

tip losses: due to the fact that $c_A \to 0$ as $r \to R$, leading to an effective shortening of the blades.

wake losses: due to the fact that the streaming air is deflected radially by the rotor blade, so that behind the rotor $\mathbf{v}_2$ is not parallel to $\mathbf{v}_1$ anymore.

5.5.1 Optimized Radial Profile of a Lift-Type Blade

In order to minimize friction losses and to maximize the extracted power, the wing breadth $b(r)$ has to be optimized as a function of the distance r from the center of the rotor. For a first approximation, we consider the schematic drawing in Fig. 5.17.

Assuming that the blade is operated close to the optimum Betz efficiency, the following relations hold:

$$v_2 = \frac{1}{3}v_1; \quad v_R = \frac{1}{2}(v_1 + v_2) = \frac{2}{3}v_1; \qquad \tan(90° - \gamma) = \frac{\frac{2}{3}v_1}{\omega r} = \frac{2}{3}\frac{R}{\lambda_S r}$$

The resulting differential lift and friction forces acting on the blade segment dr are:

$$\mathrm{d}F_A = \frac{\rho}{2}v_A^2 c_A(r)b(r)\mathrm{d}r$$
$$\mathrm{d}F_W = \frac{\rho}{2}v_A^2 c_W(r)t(r)\mathrm{d}r$$

The extracted rotor power according to the Betz approximation is:

$$P_R = c_P \cdot P_1 = \frac{16}{27}\frac{\rho}{2}v_1^3\pi R^2$$

The corresponding differential power is:

$$\mathrm{d}P_R = \frac{16}{27}\frac{\rho}{2}v_1^3 2\pi r \mathrm{d}r$$

To extract this power with z blades with the breadth $b(r)$ to be determined we need:

$$\begin{aligned}
&\mathrm{d}P = zF_A \sin(90^\circ - \gamma)\omega r \mathrm{d}r = \mathrm{d}P_R \\
&\Rightarrow z\frac{\rho}{2}v_A^2(r)c_A(r)b(r)\sin(90^\circ - \gamma)\omega r \mathrm{d}r = \frac{16}{27}\frac{\rho}{2}v_1^3 2\pi r \mathrm{d}r \\
&\Rightarrow b(r) = \frac{1}{z}\frac{16}{27}\frac{2\pi r}{c_A(r)}\frac{v_1^3}{v_A^2 \omega r \sin(90^\circ - \gamma)}
\end{aligned}$$

By introducing the tip speed ratio $\lambda_S = \frac{\omega R}{v_1}$, this can also be written as:

$$\Rightarrow b(r) = \frac{1}{z}2\pi R \frac{8}{9c_A(r)}\left(\lambda_S\sqrt{\lambda_S^2\left(\frac{r}{R}\right)^2 + \frac{4}{9}}\right)^{-1}$$

For fast rotors ($\lambda_S > 3$) and $r/R \geq 0.1$, this relation can be approximated by:

$$b(r) \approx \frac{1}{z}2\pi R \frac{8}{9c_A(r)}\frac{R}{r \cdot \lambda_S^2} \tag{5.5.1}$$

This means, that $b(r)$ should decrease hyperbolically like $1/r$ along the blade and scale with the inverse square of the tip speed ratio.

5.5.2 Losses

Taking into account the friction losses of real profiles via $c_W(r) > 0$, the real differential power becomes:

$$dP_{real} = z\omega r \left(F_A \sin(90^\circ - \gamma) - F_W \cos(90^\circ - \gamma)\right) \mathrm{d}r$$

The ratio dP_{real}/dP_{ideal} is called the "efficiency" of the profile used:

$$\eta_{profile} = 1 - \frac{F_W}{F_A \tan(90^\circ - \gamma)}$$

With the glide number $G = F_A/F_W$ and $\tan(90^\circ - \gamma) = \frac{2}{3}R/(r\lambda_S)$ this gives:

$$\eta_{profile} = 1 - \frac{1}{G\tan(90^\circ - \gamma)} = 1 - \underbrace{\frac{3}{2}\frac{r}{R}\frac{\lambda_S}{G}}_{\text{profile losses}} \tag{5.5.2}$$

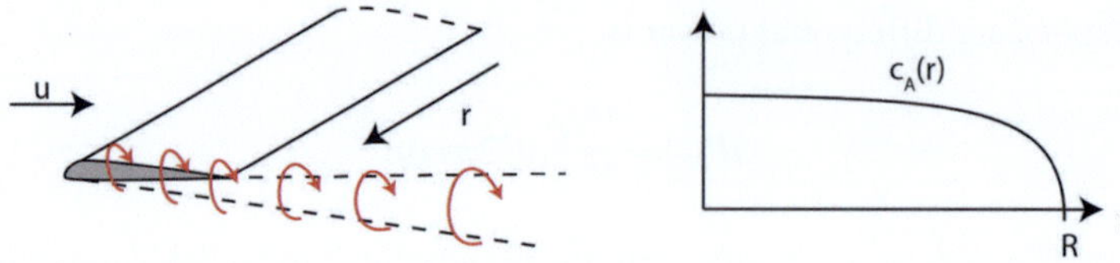

Fig. 5.18 Close to the blade tips, air can flow around the rotor plane without contributing to the lift force. As a consequence, the effective lift coefficient c_A decreases towards the end of a blade

Profile Losses

The friction or profile losses increase towards the blade tip ($r \to R$) and are proportional to λ_S. If for a given profile G is independent of r, the integration over the entire blade gives:

$$P_R = \frac{16}{27}\frac{\rho}{2}v_1^3 \int_0^R \eta_{profile} 2\pi r \, \mathrm{d}r = \frac{16}{27}\frac{\rho}{2}v_1^3 \pi R^2 \left(1 - \frac{\lambda_S}{G}\right)$$

Tip Losses

These losses[6] are due to the pressure gradient across the rotor plane. This causes air to flow radially around the blade tips, as is shown in Fig. 5.18. As a result, $c_A(r)$ decreases close to the tips. An empirical formula for tip losses of fast rotors with $\lambda_S > 3$ was provided by Betz:

$$\eta_{tip} = 1 - \frac{1.84}{z\lambda_S}$$

Wake Losses

These are caused by the forces of the blades acting on the streaming air (actio = reactio). The rotating blades give the passing air also a radial velocity component perpendicular to the direction of the incoming wind. Thus, behind the rotor plane, outgoing air follows a spiraling motion. This reduces the maximum efficiency drastically, especially for slow rotors (Fig. 5.19).

The derivation of wake losses as a function of the tip speed ratio was provided by Schmitz [3]. According to his lengthy calculation, the maximum power that can be extracted from a wind engine with wake losses is given by:

$$P_{Schmitz} = \frac{\rho}{2}v_1^3 \pi R^2 \underbrace{\int_0^1 4\lambda_S \left(\frac{r}{R}\right)^2 \frac{\sin^3(\frac{2}{3}\alpha_1)}{\sin^2(\alpha_1)} \mathrm{d}\left(\frac{r}{R}\right)}_{c_{P,Schmitz}}$$

with $\alpha_1 = \arctan\left(\frac{R}{\lambda_S r}\right)$. The integral constitutes the Schmitz efficiency, $c_{P,Schmitz}$, as a function of the tip speed ratio. In addition, the overall Betz efficiency limits

[6] Tip losses are not due to turbulences at the tips. Turbulence losses are taken into account by the profile losses.

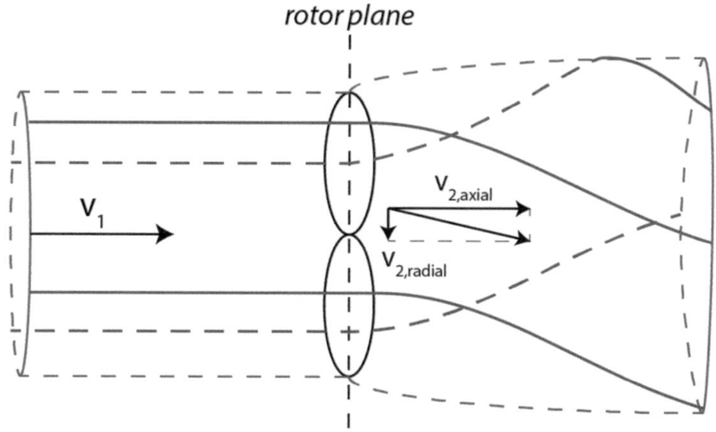

Fig. 5.19 Illustration of the origin of wake losses due to a radial velocity component $v_{2,radial}$ behind the rotor plane

Fig. 5.20 Maximum efficiency of wind engines including wake losses according to Schmitz. The wake losses are indicated by the hashed area

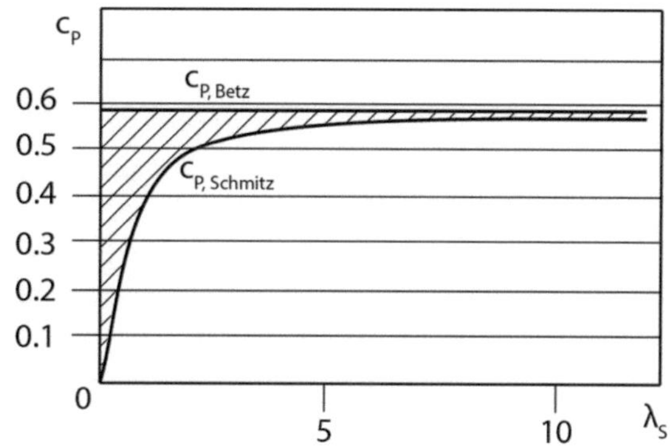

applies. Thus, as shown in Fig. 5.20, the overall efficiency of wind engines including wake losses but neglecting friction losses approaches $c_{P,Betz}$ asymptotically for $\lambda_S \to \infty$ and goes to 0 for $\lambda_S \to 0$.

5.6 Some Practical Aspects of Wind Engines

Finally, we briefly want to discuss some other practical aspects of modern wind engines. Table 5.2 summarizes typical dimensions of engines in the MW design power range. Note that most modern engines have a horizontal axis (horizontal axis wind turbine (HAWT)) and use three blades.

The reason for this becomes clear when considering real efficiencies in the frame of the efficiency limits set by Betz and Schmitz as shown in Fig. 5.21. Slow rotors are subject to large wake losses and were only used historically due to limited mechanical stability of construction materials. The best efficiencies with a maximum close to 50% are currently reached with fast rotor 3-blade HAWTs operating around tip speed

Table 5.2 Typical dimensions of various HAWT systems, depending on their nominal power

Type	Power [MW]	Height [m]	Rotor Ø [m]	Rotor area [m^2]
3 bladed HAWT	0.5	50	40	1300
	1	70	55	2400
	2	90	90	6400
	3	100	100	7900
V236-15.0MW™ [5]	8	138	236	43742

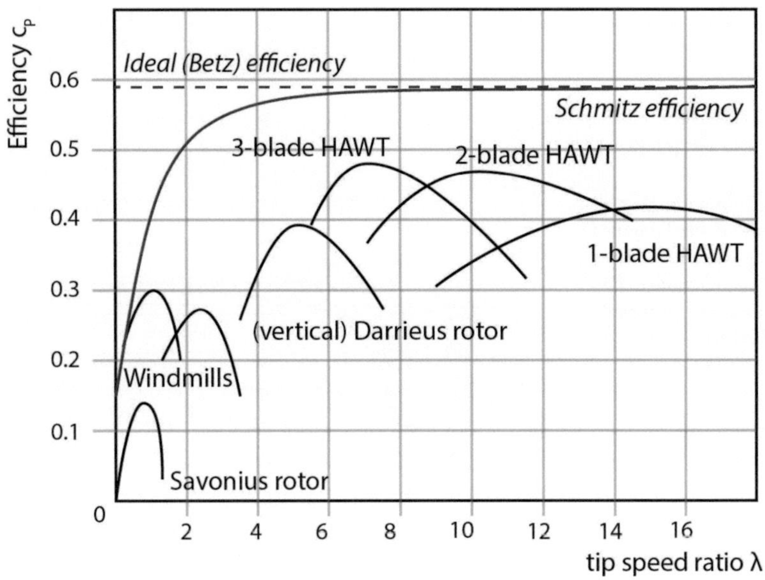

Fig. 5.21 Typical efficiencies of historical and present wind turbines as a function of the tip speed ratio. The ideal Betz and Schmitz efficiency limits are indicated for comparison

ratios of $\lambda_S = 7$. The optimum number of three blades is a result of the analysis of the optimum area coverage versus blade number and tip speed ratio discussed above. For fewer blades, the turbines have to be operated at higher tip speed ratios, which results in higher friction losses, since the friction force increases with the square of the velocity of attack.

A frequent point of concern about wind turbines is the acoustic noise generated by them. This has to be taken into account particularly when planning wind parks close to populated areas. Figure 5.22 shows how the sound intensity (measured in dB(A), decibel acoustic) emitted by a turbine decreases with distance from the rotor position. For comparison, a sound level of 100 dB(A) in close proximity would be characteristic for a very noisy industrial area. In contrast, the noise level of 40 dB(A) in a distance of 1 km corresponds to a very quiet recreational area.

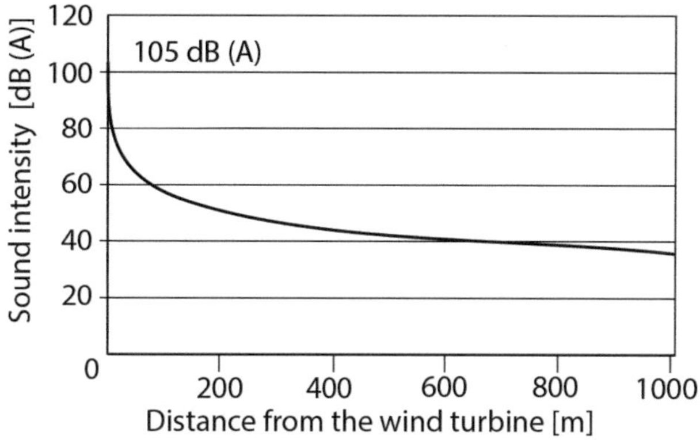

Fig. 5.22 Sound level of a typical modern wind turbine in dB(A) as a function of the distance to the rotor

Further Reading

- Time and spatially resolved animated world climate map (temperature, winds, currents, precipitation, particulates, etc.) with data from meteorological satellites: https://earth.nullschool.net/
- Bladeless oscillating column wind engine: vortexbladeless.com
- On shore wind energy kite system: skysails-power.com.

References

1. Map obtained from the Global Wind Atlas 3.1, a free, web-based application developed, owned and operated by the Technical University of Denmark (DTU). The Global Wind Atlas 3.1 is released in partnership with the World Bank Group, utilizing data provided by Vortex, using funding provided by the Energy Sector Management Assistance Program (ESMAP). For additional information. https://globalwindatlas.info
2. Betz, Albert: Das Maximum der theoretisch möglichen Ausnützung des Windes durch Windmotoren. Zeitschrift für das gesamte Turbinenwesen **26**, 307–309 (1920)
3. Schmitz, G.: Theorie und Entwurf von Windrädern optimaler Leistung, Wiss. Zeitschrift der Universität Rostock, 5. Jahrgang (1955/56)
4. Ahrens, U., Diehl, M., Schmel, R. (eds.): Airborne Wind Energy. Springer (2013)
5. Prototype 15 MW wind turbine from Vestas (DK): https://vestas.com. Accessed from 27 July 2022

Thermal Energy

6

Abstract

This chapter gives a short introduction to the basics of geothermal energy and how this potential can be exploited. The physics of heat pumps are explained as well in this sub-chapter. The second part deals with the principles of solar (thermal) energy conversion, including the physical principles of solar irradiation, its usage and some examples.

Thermal energy is generally available as waste heat in exothermal processes or can be harvested from the Earth (geothermal energy) or the radiation of the Sun (solar thermal energy). Depending on the temperature level of the heat source, thermal energy can be used directly or in combination with a heat pump for heating purposes or as process heat. Moreover, it can be transformed into e.g. electrical energy via turbine-driven generators or thermoelectric generators (discussed in Chap. 9). It also can be transported or stored on an industrial scale. Thus, thermal energy constitutes a ubiquitous and very versatile form of energy.

6.1 Geothermal Energy

6.1.1 Contributions to Geothermal Energy

The Earth was formed about 4.5 billion years ago by a process called *differential accretion* of solar dust: elements with a high density (e.g. Fe, Ni) are dominant in the center, less dense minerals (SiO_2, $Al2O_3$, CaO, MgO) are dominant in the outer mantle. The present structure of the earth is mainly known from the analysis of earthquake waves and numerical modelling and is summarized in Table 6.1.

Some points to note:

- The inner core temperature is higher than the surface temperature of the Sun. Yet the core is solid due to the high pressure. The core contains the following elements:

M. Stutzmann and C. Csoklich, *The Physics of Renewable Energy*, Graduate Texts in Physics,
https://doi.org/10.1007/978-3-031-17724-8_6

Table 6.1 Structure of the Earth

	Depth [km]	Density [tm^{-3}]	Pressure [kbar]	Temperature [K]	Phase
Inner core	5100–6370	13.5	3700	≈7000	Solid
Outer core	2900–5100	10–12	3340	≈3000	Liquid
Lower mantle	700–2900	4–5	1350	≈2000	Liquid
Transition zone	400–700	4	300	1300–2000	Viscous
Upper mantle	35–400	3–4	260	600–1300	Viscous
Crust	0–35	2.8–3.0	9	300–600	Solid

80% Fe, 7% Si, 5% Ni and 4% O. It rotates by about 0.3...0.5° / year faster than the mantle. This is called super-rotation and results in about one additional turn in ≈ 1000 years).

- The liquid outer core is responsible for the earth magnetic field.
- The Earth's crust consists of an oceanic crust which is rather thin (5–10 km) and a thicker continental crust (30–60 km). Both swim on the so-called *asthenosphere*. The crust consists to 50% of O.
- The deepest drill holes so far are: 12 km (Kola peninsula, Russia); 10 km (Oberpfalz, Germany).

The average thermal energy flow through the Earth's surface amounts to only about 60 mW m^{-2}. This is very small compared to on the average several 100W m^{-2} arriving on the earth surface due to solar irradiation. However, in local hot spots the geothermal energy flow can be higher by many orders of magnitude.

The geothermal flow has two components:

1. 50% come from cooling of the inner Earth: during accretion of the earth, the dissipated gravitational potential and kinetic energy was sufficient to reach an average temperature of 2000 K. Due to the small thermal conductivity of the crust (≈2–3 W m^{-1} K^{-1}), cooling due to radiation stopped very quickly and a stable crust developed. The geothermal flux is caused by a stable temperature gradient[1] :

$$\begin{aligned} &\text{heat flux} && \dot{Q} = 0.06\,\mathrm{Wm}^{-2} \\ &\text{thermal conductivity} && \lambda = 2\,\mathrm{W/mK} \\ &\text{temperature gradient} && \frac{\Delta T}{\Delta x} = \frac{\dot{Q}}{\lambda} \approx 0,003\,\mathrm{K/m} = 3\,\mathrm{K/km} \end{aligned}$$

   Strong deviations from the average value occur close to sites where liquid magma comes close to the surface: the resulting geothermal anomalies can have local heat fluxes up to 0.3W m^{-2} or higher.

[1] The so-called *geothermal gradient*, e.g. known from deep mines and drill holes.

Table 6.2 Most frequent radioactive decays in the Earth

Element	Decay	$T_{1/2}$ [a]	Energy [MeV]
^{235}U	α	7×10^8	≈ 100
^{238}U	α	4.5×10^9	≈ 40
^{232}Th	α	1.4×10^{10}	≈ 50
^{40}K	β^-, β^+	1.3×10^9	≈ 1.5

2. The second 50% of geothermal heat are due to radioactive decays: For an average mass concentration of radioactive isotopes of 10g t^{-1} stone (10ppm) this gives:

$$\dot{Q} = 10\,\text{g/t}\frac{6 \times 10^{23}\text{atoms}}{230\,\text{g}}\frac{1}{5 \times 10^9\text{a}} \cdot 50\,\text{MeV} \approx 10\mu\text{W t}^{-1} \approx 3\mu\text{W m}^{-3}$$

The total power reaching the Earth surface due to radioactive decays is about 16 TW, corresponding to an average flux of approximately 0.03W m^{-2}) (Table 6.2 and Fig. 6.1).

- In the young Earth the radioactive contribution was three times higher.
- 99% of the Earth are hotter than 1000°C. From this, the total thermal energy stored in the Earth can be estimated to about 3×10^{15}TW h (For comparison: the rotational energy of the Earth is about 6×13TW h).
- Strictly speaking, geothermal energy is not renewable.

6.1.2 Use of Geothermal Energy

In 2020 about 50 GW of direct geothermal energy (heating, process heat) and 15 GW for production of electricity were used, mainly in China, Sweden, USA, Iceland, Turkey, and Germany [1,3,5].

In Germany, the currently installed geothermal power for electricity generation is around 0.5 GW. In addition, about half a million heat pumps are in use, mainly for the heating of buildings.[2] In addition to heat pumps, one distinguishes the following sources of geothermal heat:

- high enthalpy sites close to geothermal anomalies
- low enthalpy sites, e.g. natural aquifers
- artificial hot dry rock (HDR) sites

[2] According to German law, geothermal heat is a "bergfreier Bodenschatz": if you own a piece of land, you do not own the heat underneath (§3 BergG).

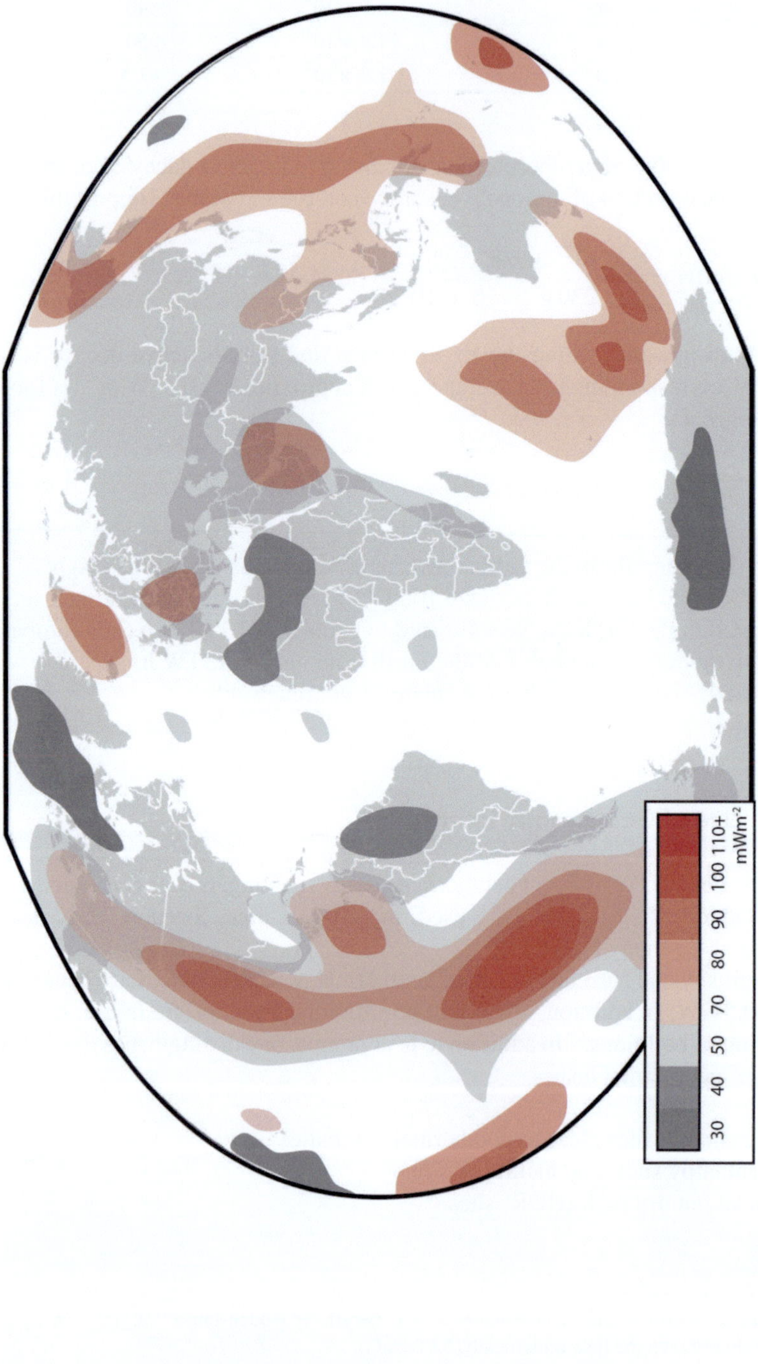

Fig. 6.1 Schematic map of the global heat flow (from the crust/surface). The variation of the heat flow ranges from ca. 30mW m^{-2} in Antarctica, to above 110mW m^{-2} in the pacific ocean at locations of the pacific fire ring. More accurate data and representations can be found at the given resources [2,3]

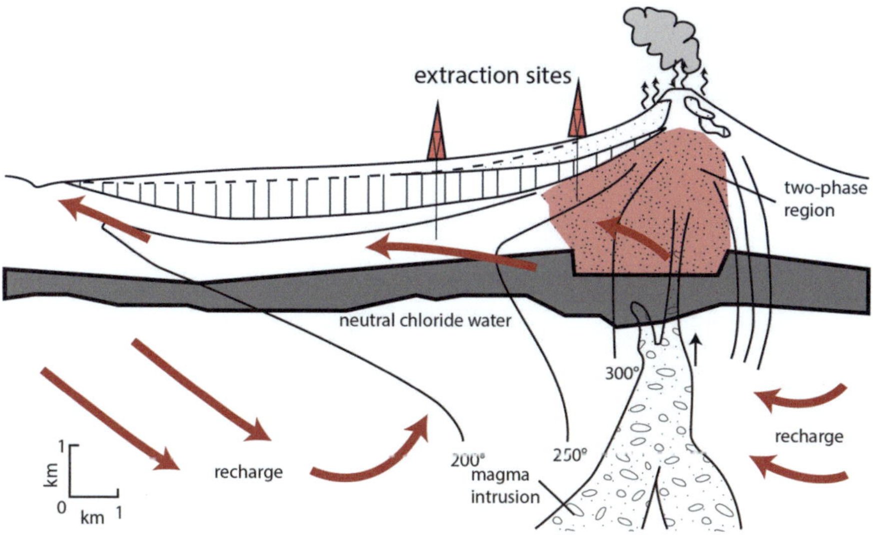

Fig. 6.2 Schematic view of a high enthalpy site close to a magma duct. Red arrows indicate main water flows. Typical lateral temperature profiles are also indicated

High enthalpy sites can be found close to active or extinct volcanoes or at other special hot spots, where one can extract water/steam mixtures (2-phase mixtures) with temperatures between 100 and 300°C, which can be used directly for electricity production with turbines. Hot 2-phase mixtures exist at high pressures (10–100 bar), so that the boiling point of water increases to about 200–300°C. Some typical properties of such high enthalpy sites are listed below:

- To extract a thermal power of 1 MW using water at 150°C one needs a volume flow $\dot{V}$ of about 2,5 Ls^{-1}.

$$\dot{Q} = \Delta T \cdot c_V \cdot \dot{V}$$

  with $c_V \approx 4\,\mathrm{kJL}^{-1}\mathrm{K}^{-1}$, and lowering the water temperature by $\Delta T = 100\,\mathrm{K}$.
- The 2-phase mixture very often is contaminated by sulphur and other minerals, giving rise to unwanted corrosion and smell. This can be avoided by using a heat exchanger, separating the geothermal water from the turbine circuit. Ideally, the resulting cold water is pumped back into the site.
- Even for extinct volcanoes there can still exist magma ducts that remain hot for several million years.
- Typical conditions at high enthalpy sites are shown schematically in Fig. 6.2.

European high enthalpy sites for electricity production are currently exploited mainly in Turkey, Southern Italy and Iceland, with a total installed power of about 3 GW [1].

Table 6.3 Low enthalpy sites in Germany

Region	Max. extractable power (GW)	For a duration of (years)
Norddeutsches Becken (Northern German basin)	100	100
Oberrheingraben (Upper rhine valley)	20	100
Süddeutsches Molassebecken (Southern German molassis basin)	30	100

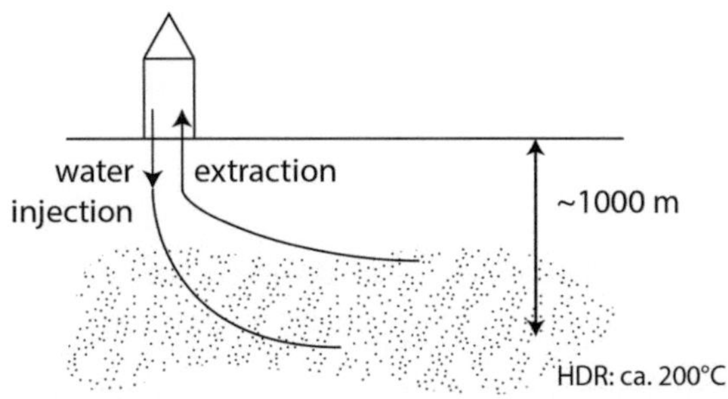

Fig. 6.3 Schematic layout of a hot dry rock (HDR) geothermal site

Low enthalpy sites typically make use of deep (1–2 km) aquifers with temperatures around 100°C. Here, the extracted water can quickly be replenished because of the high hydraulic conductivity in such aquifers.

In Germany there are three main regions for low enthalpy geothermal energy, as listed in Table 6.3.

One example for a very successful low enthalpy site is the geothermal plant in Erding, Bavaria, which extracts hot water with a relative low temperature of 65°C from the Southern German molassis basin in a depth of about 2400 m with a flow rate of 50 Ls^{-1}, generating a geothermal power of 35 MW. This is used to power the currently largest thermal spa in the world ("Therme Erding") and for the heating of residential buildings.

In regions where no aquifers can be used, one can alternatively apply the **hot-dry-rock method** (HDR) (Fig. 6.3). Here, one creates an artificial aquifer by fracturing hot rock and injecting water to extract the stored heat. As an example, granite has a heat capacity of $c_V = 840\,\text{J kg}^{-1} K^{-1}$ and a density of $2800\,\text{kg m}^{-3}$. Starting from a temperature of 200°C and cooling the extracted hot water to room temperature, this would correspond to a specific energy density of 3 g CE per kg of rock.

A major problem of HDR plants is the small thermal conductivity of the unfractured surrounding rock. It is much smaller than the thermal conductivity of aquifers, which is determined by the hydraulic conductivity according to the *law of Darcy*: if (for vertical flow) v is the flow velocity, σ_W the hydraulic conductivity, H the height of the water column above the flow region driving the flow and L the length of the flow region, then the law of Darcy states:

Table 6.4 Porosities and hydraulic conductivities of different materials

Mineral	Porosity [%]	σ_W [m/day]
Clay	45–60	$< 10^{-2}$
Sand	30–40	1–500
Gravel	25–35	500–10 000
Granite	10^{-4}	$< 10^{-3}$

$$v = \sigma_W \frac{H}{L} \tag{6.1.1}$$

Together with the specific heat of the hot flowing water, this then results in the effective thermal conductivity of the aquifer. The hydraulic conductivity σ_W is mainly, but not only, determined by the porosity of the aquifer region. Some examples are given in Table 6.4.

In the dry rock surrounding the HDR region, the thermal conductivity in contrast is determined by the transport via atomic oscillations (*phonons*) and in general is much lower. As a result, according to the estimate below, it would take very long times for surrounding heat to flow back into the HDR region.

Consider a granite cavern with a volume of $V = 100 \times 100 \times 100\,\text{m}^3$ at a starting temperature of $T = 220°\text{C}$. For a density of $3000\,\text{kg m}^{-3}$, the total rock mass in the cavern is $m = 3 \times 10^9\text{kg}$.

- The total energy content of the HDR region in the cavern (when cooling down to 20 C) is:

$$Q = T \cdot m \cdot c_V = 200\,\text{K} \cdot 3 \times 10^9\,\text{kg} \cdot 840\,\text{J kg}^{-1}\text{K}^{-1} = 5 \times 10^{14}\text{J}$$

- For extracting a heat flow of $\dot{Q} = 10\,\text{MW}$, this reservoir will last for about one year
- The heat transport from the surrounding back to the cavern is determined by the normal thermal conductivity λ:

$$\dot{Q} = \lambda \Delta T \frac{A}{d}$$

With a surface area of the cavern of $A = 6 \times 10^4\text{m}^2$, assuming a temperature gradient of $\Delta T/d = 1\,\text{Km}^{-1}$ and the typical low thermal conductivity of solid rock, it would take about 100 years to refill the extracted heat.

Fig. 6.4 Working principle of a compression heat pump

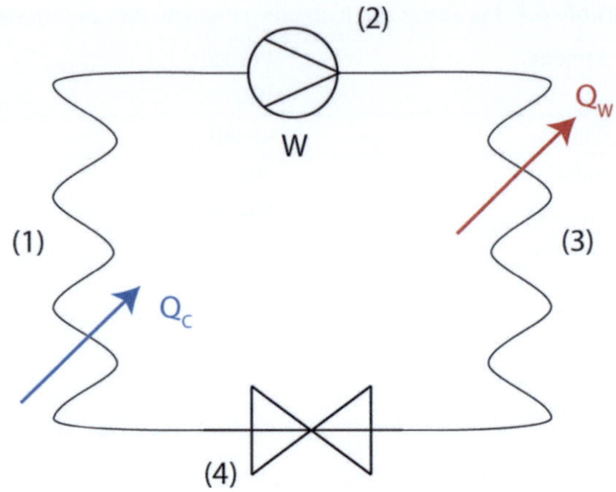

If the temperature of the extracted geothermal water is not sufficiently high for the intended use, the temperature is usually increased with the help of a heat pump. **Heat pumps** are heat engines running in reverse. They convert work into heat and at the same time pump additional heat from a cold to a warm reservoir (c.f. Fig. 1.2 in Chap. 1). Heat Q_C is extracted from a cold reservoir (at temperature T_C) and pumped with the help of external work W into a warm reservoir (Q_W at temperature T_W).

Energy conservation requires that $Q_W = Q_C + W$. Therefore, the ideal thermodynamic efficiency of a heat pump is given by:

$$\boxed{\eta_{hp} = \frac{Q_W}{W} = \frac{T_W}{T_W - T_C} > 1}$$

The main technical components of a compression heat pump are shown in the schematic diagram of Fig. 6.4.

1. The working medium is evaporated in a heat exchanger and thereby removes the necessary heat of evaporation Q_c from the cold reservoir.
2. The now gaseous medium is compressed in a compressor, thereby adding the mechanical work W (as adiabatically as possible).
3. A second heat exchanger transfers the heat Q_W to the warm reservoir, causing liquefaction of the working medium.
4. A throttle between the two heat exchangers is used to adjust the different pressure levels on both sides

The most common reservoirs for T_C are outside air, groundwater or deeper surface regions in the earth, where seasonal variations of the temperature are strongly reduced. Some typical working media for heat pumps are listed in Table 6.5.

As a practical example, consider a floor heating system with a temperature $T_W = 50°C$ using groundwater with $T_C = 5°C$ as the cold reservoir. Transforming degrees

Table 6.5 Common working media for heat pumps

Medium	Boiling point °C
CCl_2F_2 (no longer legal)	–30
C_3H_8 (propane)	–42
NH_3	–33
CO_2	–57

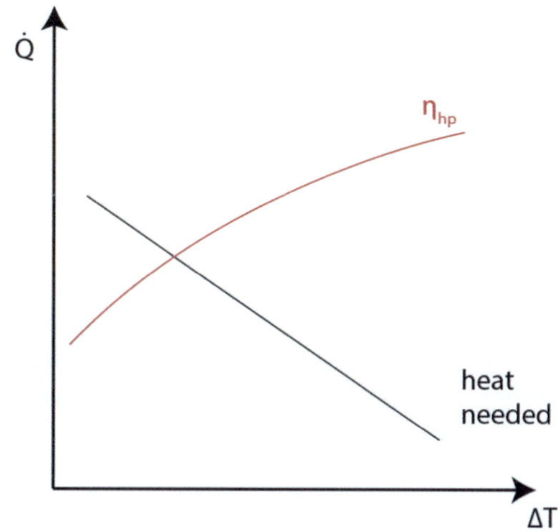

Fig. 6.5 Qualitative seasonal divergence between the heat needed and heat pump efficiency for building heating with a heat pump using outside air as the cold reservoir: in the winter, T_C is low, so the temperature spread $\Delta T = T_W - T_C$ is large. This causes a lower efficiency, although the required heat flux is large. In the summer, T_C approaches T_W, so the efficiency increases, but much less heat is required

Celsius into absolute temperatures T, this would be possible with an ideal efficiency of:

$$\eta_{hp,ideal} = \frac{50 + 273}{50 - 5} \approx 7 \tag{6.1.2}$$

Typical efficiencies of commercial heat pumps are between 2 and 6, depending on the temperature spread $T_W - T_C$. This, for example, gives rise to a seasonal divergence between the heating power of a heat pump and the amount of heat needed in a building (Fig. 6.5).

The overall system efficiency η_{system} of a heat pump is lower than η_{hp}, because the work W can only be provided with a finite efficiency η_W:

$$\eta_{system} = \eta_{hp} \cdot \eta_W$$

Real heat pumps have a lower efficiency of $\eta_{hp,real} \approx 1/2 \cdot \eta_{Hp,ideal}$. If the external work is provided by an electrically driven compressor, as it is commonly the case, then also $\eta_W \approx 0.35$, limited by the overall efficiency of commercial electrical power plants. Then, the resulting system efficiency is $\eta_{system} \approx 0.35 \cdot 3.5 \approx 1$. This means that, viewed on a system-wide scale, the use of a heat pump only recovers the energy lost during electric power generation in a thermal power plant. A better solution is provided by the use of a combustion engine for W ($\eta_W \approx 0.25$). If in addition the waste heat of the combustion engine is used to increase T_C (e.g. 5°C → 35°C), then the overall efficiency would be increased to:

$$\eta_{system} = 0.25 \cdot 1/2 \frac{50 + 273}{15} \approx 3 \tag{6.1.3}$$

6.2 Solar Thermal Energy

The thermal use of solar radiation requires the efficient transformation of the broad spectrum of solar photons impinging on the Earth into heat in the form of low energy thermal vibrations (*phonons*) in the respective absorber materials. Only in the small infrared portion of the solar spectrum, this transformation can occur directly through the absorption of infrared photons by excitation of atomic vibrational modes in molecules or solids. For the much larger part of the spectrum, the photons first excite electrons into higher energy states, which subsequently decay back to the electronic ground state by the generation of atomic vibrations. This process is called *thermalization* and is governed by the details of the coupling between electrons and phonons in the absorber material.

As the simplest example, we briefly discuss the electron-phonon coupling in molecules.[3] As shown in the sketches below, in isolated atoms photons can couple to electrons only via transitions between the discrete energy levels of the atomic orbitals driven by the oscillating electric field of the photons. In these processes, for photon energies in the eV range, the atoms basically remain at rest because of the very small momentum of the photons. In molecules, however, the situation is quite different. Here, an optically excited electronic transition e.g. from the molecular ground state (bonding orbital) to an excited antibonding orbital will lead to a rearrangement of the equilibrium distance between the two atoms forming the bond. The very fast electronic transition into an antibonding level causes a sudden repulsion of the atoms, which excites a harmonic vibration with frequency Ω around the new equilibrium distance due to the inertia of the atomic nuclei (Fig. 6.6).

If many molecules can interact with each other in condensed matter (gases, liquids, solids), the excitation of a given molecule will spread to all other molecules, until thermal equilibrium among all possible vibrational (and rotational) modes is reached, resulting in a corresponding increase in temperature.

There are four basic processes through which photons interact with condensed matter:

[3] (More details are provided in Chap. 8).

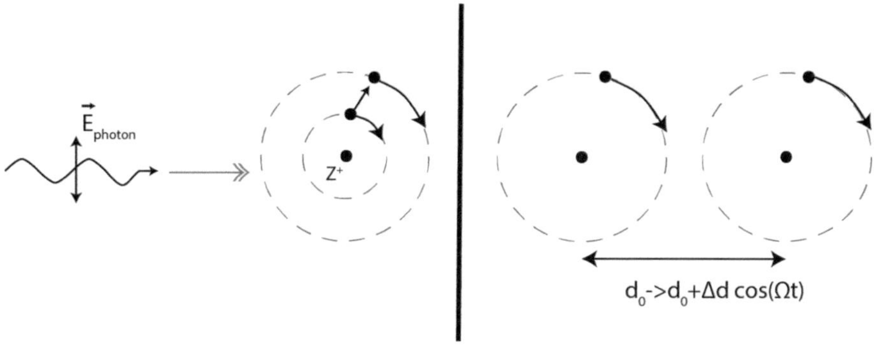

Fig. 6.6 Photon absorption in atoms only increases electronic energy. In molecules or solids, part of the electronic excitation gives rise to molecular vibrations via the electron-phonon coupling

1. **Reflection, reflectivity R**: This is a coherent process at an interface, depending on the indices of refraction on both sides, the polarization of the light, and the angle of incidence. Details are described by the general *Fresnel formulae.*
2. **Scattering S**: An incoherent process due to the interaction with atoms, molecules, or small particles. *Rayleigh scattering* occurs when the wavelength of light λ is much larger than the size of the scattering object L. It is isotropic and increases with decreasing λ. *Mie scattering* occurs at small particles or interface roughness for $\lambda << L$
3. **Absorption, absorbance A**: This is an incoherent process due to the transformation of photons to excited electrons and/or phonons. In homogeneous media the *Lambert-Beer law* holds for the local light intensity $\Phi(x)$ inside the medium ($\alpha = \alpha(\hbar\omega)$ is the *absorption coefficient*):

$$\Phi(x) = (1 - R)\Phi(0)\exp(-\alpha x) \tag{6.2.1}$$

4. **Transmission T**: Fraction of the incoming photon flux passing (transmitted) through the medium.

Conservation of energy yields:

$$\boxed{R + S + A + T = 1} \tag{6.2.2}$$

For an optimum harvesting of solar thermal energy, all processes need to be optimized. This is achieved by technically approaching the following idealized optical elements:

- ideal absorbers (A = 1 for all wavelengths)
- selective absorbers (A = 1 in one spectral range, A = 0 for the rest)
- ideal mirrors (R = 1, e.g. for concentrating solar receivers)
- ideal anti-reflection coatings (R = 0, compare Chap. 8)
- scattering in the environment, on surfaces and in inhomogeneous media

Table 6.6 Global average values of atmospheric absorption, scattering and transmittance

γ_S	AM	A [%]	$S_{Rayleigh}$ [%]	S_{Mie} [%]	T [%]
90°	1.0	8.7	9.4	0–25	62–83
60°	1.15	9.2	10.5	1–30	57–81
30°	2.0	11.2	16.3	4–45	41–71
5°	11.5	19.5	42.5	25–86	6–35

Table 6.7 Seasonal variations in Berlin (ca. 52° N, varying γ_S, 10 a average)

Date	γ_S	AM	$\hat{\Phi}_{direct}$ [$k\,W\,h\,m^{-2}d^{-1}$]	$\hat{\Phi}_{diffuse}$ [$k\,W\,h\,m^{-2}d^{-}$]
30.1.	19.5°	3	0.17	0.44
1.4.	42°	1.5	1.42	2.07
21.6.	61°	1.15	2.58	2.86
22.12.	14°	4	0.1	0.35

Table 6.8 Local variations of these values (yearly average) in Europe

Site	Latitude (N)	$\hat{\Phi}_{direct}$ [$k\,W\,hm^{-2}d^{-1}$]	$\hat{\Phi}_{diffuse}$ [$k\,W\,hm^{-2}d^{-1}$]
Bergen	60°	0.86	1.29
Vienna	48°	1.4	1.63
Lisbon	38°	3.06	1.66

In addition, condensed matter emits radiation also without incoming photons at finite temperatures according to the Stefan-Boltzmann law, depending on the emissivity e (cf. Sect. 3.1).

Important for solar thermal energy are first of all absorption and scattering of sunlight in the atmosphere. This strongly depends on the season and angle of incidence γ_S of sunlight relative to the surface of the earth, as exemplified in Tables 6.6 to 6.8.

All values depend strongly on the so-called *air mass (AM)*, describing the relative length of travel of sunlight through the atmosphere for different angles of incidence:

$$\boxed{AM = \frac{1}{\sin \gamma_S}} \tag{6.2.3}$$

The large percentage of scattered light leads to diffuse radiation, which can not be concentrated anymore. Average seasonal and local values for direct versus diffuse sunlight energy flux densities per day are listed in Tables 6.7 and 6.8.

Because of the strong variation of γ_S during the day, an active positioning of solar collectors improves the received energy. In central Europe, the typical gain in energy is:

- for two-axis solar trackers: $\approx 30\%$ more energy
- for one-axis solar trackers: $\approx 20\%$ more energyn

Currently, there are three main fields of applications of solar thermal energy:

1. The use of solar thermal elements in building architecture for the design of low-energy houses, such as winter gardens, intelligent ventilation systems, active windows, …
2. Non-concentrating (flat) solar collectors: these use a combination of a glass cover (selective transmission: high transmission in the visible, low transmission in the infrared spectral range) with a selective absorber (high absorbance in the visible spectral range, low emissivity in the infrared region). Since according to the Kirchhoff law of radiation the absorbance A and the emissivity e at a given wavelength λ are equal for directed radiation at thermal equilibrium, an ideal selective absorber would have the spectral characteristics shown in Fig. 6.7. Commonly used materials for such selective absorbers are "black chromium", "black nickel", or $TiNO_X$.
3. Concentrating solar receivers: These use mirrors to focus the direct sunlight onto a thermal receiver. Depending on the degree of concentration, different receiver temperatures can be achieved. This is then used to generate steam, heat up oil,

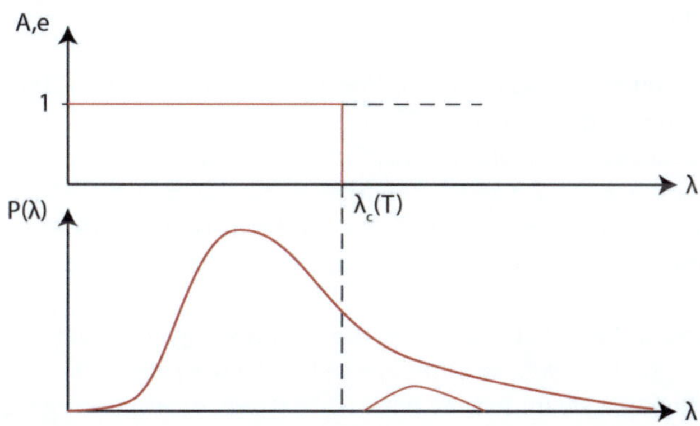

Fig. 6.7 Spectral dependence of the absorbance A and the emissivity e of an ideal selective absorber. The absorbance is high in the ultraviolet and visible region of the solar spectrum $P(\lambda)$ schematically shown in the lower panel. A and e drop to zero beyond a critical wavelength λ_c, depending on the absorber temperature T. In this way the emission of thermal radiation from the absorber is minimized

drive a Stirling motor, melt phase change salts, drive thermochemical reactions, etc.
The optical concentration or focusing of direct sunlight is limited by the finite half opening angle of the Sun as seen from the Earth (cf. Sect. 3.1):

$$\alpha = \arcsin \frac{R_\odot}{r_{SE}} = 16' = 0.27^\circ = 0.00465\,\text{rad} \tag{6.2.4}$$

For biaxial concentration e.g. with a circular parabolic mirror this results in a maximum concentration factor of $C_{max} = 1/\alpha^2 = 46211$. For a uniaxial, linear concentrator (e.g. a parabolic trough) the physical limit is $C_{max,lin} = 1/\alpha = \sqrt{C_{max}} = 215$. The maximum working temperature T_W of an insulated receiver with emissivity e=1 is then obtained from the Stefan-Boltzman equation as:

$$C \cdot SC = \sigma T_W^4 \quad \Rightarrow \quad T_W = \left(\frac{C \cdot SC}{\sigma} \right)^{1/4} \tag{6.2.5}$$

where SC is the solar constant. The maximum working temperature $T_{W,max}$ is reached for C_{max}. The temperature in the focus of an ideal concentrator would then approach the effective surface temperature of the Sun:

$$T_{W,max} = 5780\,\text{K} = T_{eff,sun}; \quad \Rightarrow \quad T_W = T_{eff,sun} \left(\frac{C}{C_{max}} \right)^{1/4} \tag{6.2.6}$$

Practical receiver temperatures of concentrating solar thermal systems are typically limited to $< 1500\,\text{K}$, due to issues with long-term receiver material stability.

Some examples of concentrating solar thermal systems are listed below:

- The “Euro trough”: a standardized, linear parabolic collector (2001). Length = 150 m, width = 5.80 m, area = 825 m^2, $C = 82$
- The “Solar Electric Generation System” (SEGS) in the Mojave dessert, California (since 1984). Area = 2.3 Mm^2, >1 Mio. parabolic troughs, 350 MW installed electric power, oil at 300–400 °C as the working medium, driving conventional steam turbines. This power plant operates at an overall efficiency of $\eta \approx 30-38\%$.
- Solar towers in large heliostat fields are operated e.g. in Spain at Almeria and Sevilla. $C \approx 600 - 1000$, air or molten salt as the working medium, $T_W \approx 700 - 1000$°C, driving a steam turbine with 10 MW electrical power generation at an overall efficiency of $\eta \approx 15\%$.
- An example of current research is shown in Fig. 6.8. It aims at the thermochemical generation of bio-fuels via the production of syngas, a mixture of H_2 and CO. To this end, the reduced form of a metal oxide (MO_{red}) is used at a lower temperature level T_C to reduce a mixture of water and CO_2, resulting in an oxygen-rich metal oxide (MO_{ox}) plus syngas in an overall exothermic reaction. MO_{ox} is then

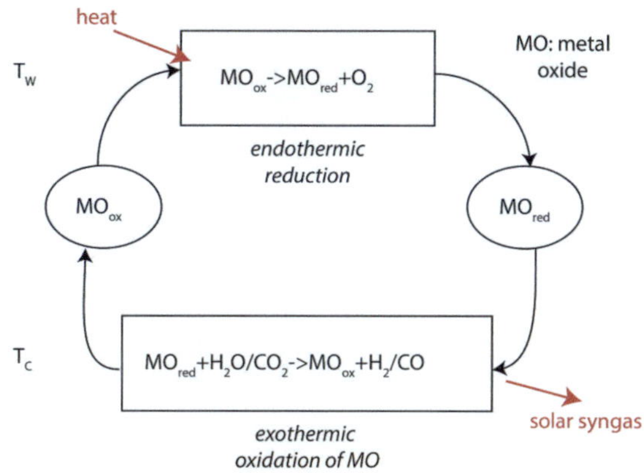

Fig. 6.8 Schematic flow diagram for solar syngas production. See text above for details

transported to a solar receiver with a working temperature T_W, which is sufficiently high to reduce MO_{ox} back to MO_{red} by splitting off oxygen in an overall endothermic reaction. Then, the cycle starts again.

Further Reading

- Hellisheiði Power Station, the largest geothermal power plant in Iceland (300MWe: https://www.on.is/en/about-us/power-plants/)
- The world's largest thermal spa in Erding is powered by the geothermal plant Erding: https://www.steag.com/de/erding, https://www.geowaerme-erding.de/.
- https://www.statista.com/statistics/476281/global-capacity-of-geothermal-energy/
- Tischner, T., Melchert, B., Ortiz, A., Schindler, M., Scheiber, J., Genter, A.: Test- und Probebetrieb des HDR-Kraftwerks Soultz. Abschlussbericht zum BMU-Vorhaben 0325159, 1–128 (2013).
- Summary of home use heat pumps: https://www.energy.gov/energysaver/heat-pump-systems.
- A recent review on selective absorbers: Zhang, J., Wang, C., Shi, J., Wei, D., Zhao, H., Ma, C.: Solar Selective Absorber for Emerging Sustainable Applications. Adv. Ener. Sust. Res., **3(3)**, 2100195 (2022)
- Fortuin, S., Stryi-Hipp, G.: Solar Collectors, Non-concentrating. In: Richter, C., Lincot, D., Gueymard, C.A. (eds.) Solar Energy. Springer, New York, NY (2013). https://doi.org/10.1007/978-1-4614-5806-7_681
- Solar tower Jülich: https://www.dlr.de/content/en/research-facilities/solar-tower-juelich.html
- SEGS: https://en.wikipedia.org/wiki/Solar_Energy_Generating_Systems
- Solar towers: https://en.wikipedia.org/wiki/Ivanpah_Solar_Power_Facility

References

1. 2020 EGEC Geothermal Market Report, Garabetian, T., et al. (eds.): © EGEC - European Geothermal Energy Council (2021)
2. Davies, J.H.: Global map of solid Earth surface heat flow. Geochem. Geophys. Geosyst. **14**(10), 4608–4622 (2013)
3. Fuchs, S., Norden, B.: International Heat Flow Commission (2021): The Global Heat Flow Database: Release 2021. GFZ Data Services (2021). https://doi.org/10.5880/fidgeo.2021.014
4. Huttrer G.W.: Geothermal Power Generation in the World 2015-2020 Update Report, Proceedings World Geothermal Congress 2020 Reykjavik (2021)
5. Analyzed data from the BP Statistical Review of World Energy on https://ourworldindata.org/grapher/installed-geothermal-capacity?tab=chart&country=USA~ISL~TUR~OWID_WRLourworldindata.org. Accessed 28 July 2022
6. Lund, J.W., Toth, A.N.: Direct Utilization of Geothermal Energy 2020 Worldwide Review. In: Proceedings World Geothermal Congress 2020 Reykjavik (2021)

Photosynthesis

7

Abstract

In this chapter the different forms of biomass and their relative and absolute energy content are discussed. The detailed biophysical mechanism of photosynthesis is explained as the main energy conversion process in plants. Also, a brief introduction to artificial photosynthesis is presented.

Photosynthesis is the natural transformation of atmospheric CO_2 into biomass through photo-catalytic water splitting and CO_2 reduction powered by sunlight. Although photosynthesis is mainly a biochemical process, it also has a significant number of physical aspects which will be considered in this chapter. After all, the products of photosynthesis in the form of biofuels and fossil fuels have been and still continue to be the main source of primary energy for mankind. In addition to N_2, CO_2 was the main component of the Earth's atmosphere during the first three billion years of Earth history. Only about one billion years ago, the oxygen partial pressure in the atmosphere for the first time exceeded its CO_2 content because of the increasing photosynthetic activity in the oceans and on the surface of the Earth. The current atmospheric composition with about 20% O_2 and less than 1000 ppm CO_2 was only reached about 300 million years ago, enabling the development of higher life forms with an oxygen-based metabolisms.

7.1 General Considerations of Biomass Usage

In view of renewable energy, the main outcome of photosynthesis is biomass. As shown in Fig. 7.1, the amount of global carbon present as biomass is about 600 Gigatons (Gt). This is very small compared to carbon stored in the soil, the ocean waters and the lithosphere. A similar amount of carbon is present in the atmosphere, mainly in the form of CO_2. Moreover, about 10 000 Gt of carbon are stored as ancient biomass in the Earth in the form of fossil fuels. Both, the surface of the sea and the land

M. Stutzmann and C. Csoklich, *The Physics of Renewable Energy*,
Graduate Texts in Physics,
https://doi.org/10.1007/978-3-031-17724-8_7

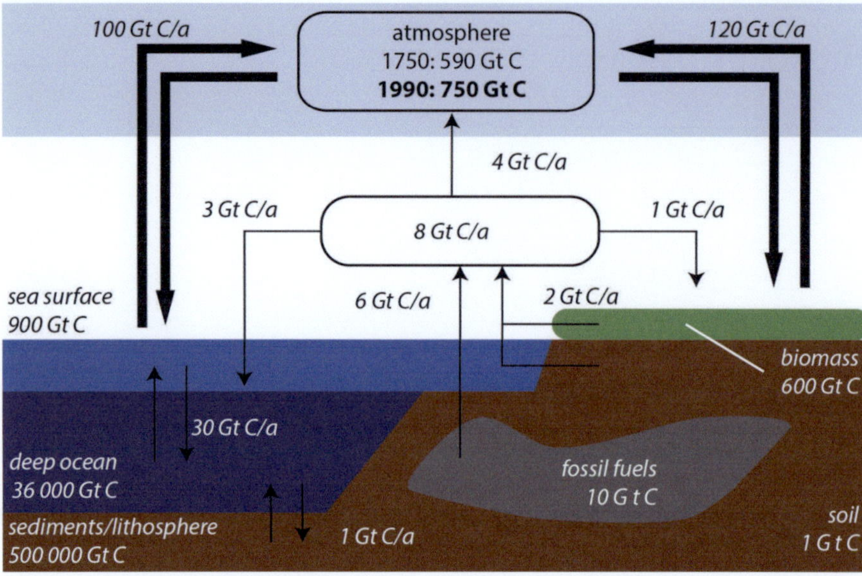

Fig. 7.1 Global carbon inventories in the atmosphere, the oceans, land and the lithosphere. Black arrows indicate annual carbon fluxes between these different reservoirs. In the center, the anthropogenic contributions to the global carbon fluxes are shown

exchange about 100 Gt of carbon annually in both directions. The human footprint in the annual carbon cycle amounts to about 8 Gt,[1] mainly coming from the use of fossil fuels and surface biomass. 4 Gt of these are absorbed back by the oceans and the land, while the other 4 Gt are released into the atmosphere, gradually increasing the amount of carbon there [1].

The transformation of CO_2 and water via the use of sunlight into biomass by plants, algae and bacteria is the most important basis for the energy presently used by mankind:

- Fossil fuels (oil, gas, coal) have been created by photosynthesis and currently provide ≈82% of the energy used [2].
- In developing countries ≈35% of all primary energy comes from surface biomass. Its use occurs mainly via low-tech fireplaces with a very low efficiency of about 15%. In addition, a lot of atmospheric pollution is caused by the resulting smoke [3].
- In Europe and the USA only around 3% of the primary energy come from surface biomass [2,3].

[1] Emission of pure C. This corresponds to 30 Gt of CO_2.

Table 7.1 Energy production by different photosynthesis systems. The corresponding values for large scale Si solar cells are shown for comparison

System	Biofuel production [kg/(ha · year)]	Energy content [kWh/(ha · year)]	Efficiency [%]
Oil palm	4000	35000	0.17
Jatropha	2500	25000	0.12
Sugar cane	2500	16000	0.08
Micro algae	90000	$1 \cdot 10^6$	4.5
Si solar cell	–	$3 \cdot 10^6$	14.0

As a typical example for the present use of surface biomass in central Europe we mention the state of Bavaria in Germany, which is a highly developed industrial region with a still considerable amount of agricultural area spanning $\approx$2 Mio. ha. Of these, about 10% are used for the production of biofuels in the form of fire wood (40%), biomass for power plants (40%), liquid biofuels (10%) and biogas (10%) [4].

A very problematic aspect of the use of surface biomass is the direct competition between biomass used for biofuels and for food, as well as the strong usage of sweet water.

The global production of biomass (per year) amounts to about:

- $1,2 \times 10^{11}$t of dry biomass on land with a total area of $1 \times 10^8\,\text{km}^2$
- 6×10^{10}t of dry biomass in the oceans with a total area of $3,6 \times 10^8\,\text{km}^2$

The average energy content of dry biomass is $\approx 20\,\text{MJ kg}^{-1}$. Thus, the global energy potential of biomass amounts to about:

$$1.8 \times 10^{14}\,\text{kg} \cdot 2 \times 10^7\,\text{MJ kg}^{-1} = 3.6 \times 10^{21}\,\text{J}$$

of total energy per year, which is actually 10 times the total energy consumption of mankind. On the other hand, compared to the total solar irradiation of 4×10^{24} J per year, this gives a global efficiency of photosynthesis of about 0.1%. As a matter of fact, most of the incoming solar energy is used to guarantee the long-term survival and progression of biomass. In comparison, as will be discussed below, the primary efficiency of the photosynthesis reaction is $\approx$30%, which constitutes the long-term aim of artificial photosynthesis discussed at the end of this Chapter.

As exemplified in Table 7.1, even for natural photosynthesis the overall efficiency can vary a lot and is easily surpassed by technical use of sunlight e.g. in the form of solar cells.

7.2 Biophysical Principles of Photosynthesis

In general, one distinguishes two different forms of photosynthesis, depending on the underlying basic chemical reactions:

Oxygen-generating photosynthesis of most green plants according to the overall reaction

$$H_2O + CO_2 \xrightarrow{\hbar\omega} (CH_2O) + O_2$$

An-oxygeneous photosynthesis of bacteria using H_2S instead of H_2O, e.g. in the vicinity of sulphur-containing vulcanic fumeroles in the oceans:

$$H_2S + CO_2 \xrightarrow{\hbar\omega} (CH_2O) + O + S$$

As an example, the overall reaction for photosynthetic production of glucose is:

$$12H_2O + 6CO_2 \xrightarrow{\hbar\omega} C_6H_{12}O_6 + 6O_2 \qquad \Delta H = 2870\,\text{kJ/mol}$$

The complete photosynthetic cycle is divided into a light reaction (primary reaction) and a dark reaction (secondary reaction, *Calvin cycle*), as shown in Fig. 7.2:

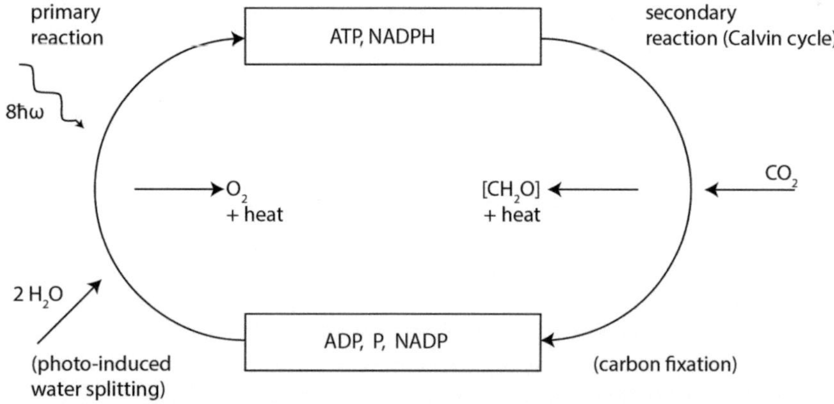

Fig. 7.2 Illustration of the two components of the oxygen-generating photosynthetic cycle: in the primary reaction, two water molecules are split into hydrogen and oxygen by the absorption of eight photons in the red spectral range, resulting in the formation of molecular oxygen and the intermediate energy carriers ATP and NADPH (see below). In the following secondary reaction in the dark, the intermediate carriers drive the fixation of carbon from CO_2 in (CH_2O), returning the intermediate carriers into their low energy forms ADP and NADP

For the fixation of one carbon atom two water molecules need to be split:

$$2H_2O \xrightarrow{8\hbar\omega} O_2 + 4H_+ + 4e^-$$

For the generation of one free electron, two photons of wavelength $\approx$700 nm ($\hbar\omega = 1,8$ eV, red) are necessary. The minimum energy to split water is $\approx$1,23 eV. Per fixated carbon atom about 4 eV of chemical energy are stored. The resulting total energy efficiency of the primary reaction is therefore:

$$\eta = \frac{4\,\text{eV}}{8 \cdot 1.8\,\text{eV}} \approx 30\%$$

Remarkably, this energy efficiency is close to the maximum thermodynamic efficiency of solar cells, the so-called *Shockley-Queisser* limit. See Chap. 8 for more details.

The protons and electrons produced by the primary reaction are used to generate the intermediate energy storage molecules adenosine diphosphate (ADP)/adenosine triphosphate (ATP) and nicotineamid adenine dinucleotid phosphate (NADP):

$$\text{ADP} + \text{P} \rightleftharpoons \text{ATP}$$
$$\text{NADP}^+ + 2e^- + \text{H}^+ \rightleftharpoons \text{NADPH}$$

The necessary reaction takes place in the *thylakoid membrane* of chloroplasts following the so-called *Z-scheme*, schematically shown in Fig. 7.3.

The photosynthetic process starts with absorption of red photons (wavelengths 680 and 700 nm) in the antenna complexes of the photosystems I and II (PS I and PS II). These antenna complexes consist out of circular arrangements of several hundred chlorophyll molecules. The main constituents of such planar chlorophyll molecules are four pyrrole rings (C_4N rings with aromatic electron structure) arranged around a central Mg atom. Optical absorption by one of the double bonds in such rings excites the chlorophyll molecules from their ground state *Chl* to their excited state *Chl**. Details of this absorption process are discussed in the next section. The ground state hole in PS II catalyzes the splitting of water into O and protons H^+. The excited electron in *Chl** is passed on rapidly to a chain of other molecules detailed in Fig. 7.3 downward in energy to separate it spatially from the ground state hole, avoiding direct recombination. Part of the freed electronic energy is used to generate ATP from ADP and P in a process called electron transfer phosphorylization (ETP). Finally, the excited electron from PS II ends up in PS I to neutralize the optically excited hole there. The excited electron of PS I is again separated from the ground state hole by a downward transfer process with the molecule ferredoxin (Fd) as an important functional ingredient. Fd stores two electrons from two subsequent optical

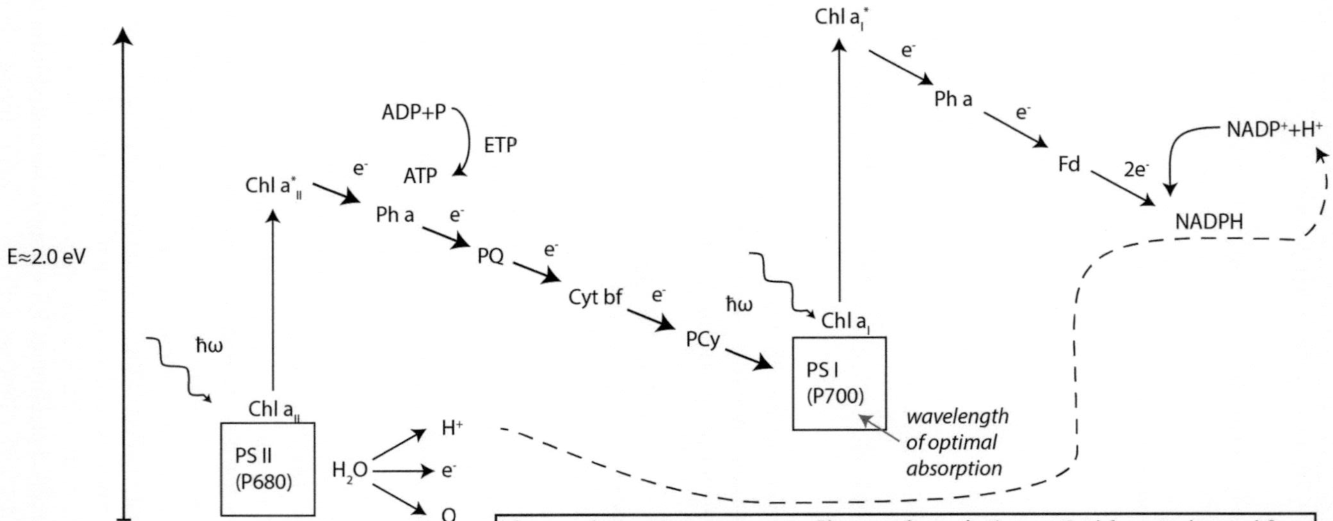

PS:	photosystems	Ph a:	phaeophytin a	Cyt bf:	cytochrome bf
Chl a:	chlorophyll a	Fd:	ferredoxin	PCy:	plastocyanin
	(a^*: excited state after optical absorption)	PQ:	plastoquinon	ETP:	electron transport phosphorylization

Fig. 7.3 Optical absorption and electron transfers according to the *Z-scheme* of photosynthesis. See text for details

excitation events of PS I, which then enable the synthesis of the intermediate energy carrier NADPH from NADP+ and H^+. Again, further molecular details are discussed in the next section.

The final fixation of carbon from CO_2 in higher carbohydrates (the Calvin cycle) is a very complicated sequence of chemical reactions and will not be discussed here in any detail. It suffices to mention that for carbon fixation one distinguishes C_3– and C_4–plants:

- C_3-plants bind ≤ 30 mg CO_2/ 100 cm^2 leaf area and hour
- C_4-plants bind 50–90 mg CO_2/ 100 cm^2 leaf area and hour

The overall efficiency of photosynthesis also depends on:

1. wavelength of light: chlorophyll a and b[2] mainly absorb in the blue (400–480 nm) and red (600–700 nm) spectral range, giving rise to the green colour of leaves. For wavelengths $\lambda > 690$ nm the photosynthesis stops abruptly, known as the "red drop" of photosynthesis. Using different pigments than chlorophyll also other spectral ranges can be used (e.g. via carotinoids, bacteriorhodopsin).
2. Light intensity: photosynthesis starts at around $10\,W/m^2$ with an optimum at approximately $230\,sW/m^2$.
3. Temperature: a broad optimum is reached at 35 °C
4. CO_2 content: the optimum concentration is at $\approx$1vol% (i.e. much higher than the 400 ppm = 0.04% of CO_2 currently in our atmosphere).

7.3 Basic Biomolecular Processes of Photosynthesis

In this section, we discuss some of the biomolecular processes mentioned above in more detail. For successful photosynthesis, four distinct sub-reactions have to be coordinated spatially, energetically and chemically: The first sub-reaction occurs in **photosystem I** (PS I) and uses two photoexcited electrons to produce twofold reduced ferredoxin (Fd^{2-}) by storing the electrons in the central Fe-S molecular orbital of ferredoxin. This is shown schematically in Fig. 7.4. The two electrons are fed into the ferredoxin via the phaeophytin a (Ph a) complex, which enables a fast removal of the photoexcited electron from the reaction center of the photosystem:

$$2e^- + Fd_{ox} \longrightarrow Fd^{2-}_{red}$$

The two electrons stored in ferredoxin then enable the synthesis of NADPH:

$$Fd^{2-}_{red} + NADP^+ + H^+ \longrightarrow Fd_{ox} + NADPH$$

[2] Chl a/b: chlorophyll a/b.

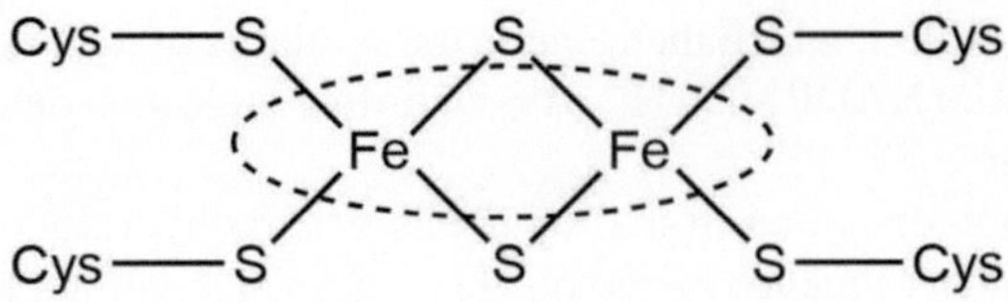

Fig. 7.4 Schematic molecular structure of ferredoxin. The area encircled by the dashed line indicates the molecular Fe–S orbital capable of storing two additional electrons. Cys symbolizes cystein end groups

This second step is catalyzed by the enzyme ferredoxin-NADP-reductase.

Figure 7.5 shows the relative energetic positions of the different molecules involved in the synthesis of NADPH.

The energy scale used is the redox potential relative to the normal hydrogen electrode (NHE)[3] potential as the origin.

The second sub-reaction takes place at the **photosystem II** (PS II). After optical excitation and transfer of the excited electron to Ph a, the remaining positively charged reaction center P680$^+$ acts as a very strong oxidant which, in combination with the oxygen-evolving center (OEC) complex Mn-O, catalyses the formation of O_2 via splitting of two water molecules

Two excited electrons from PS II and the protons from the water-splitting reaction are transferred to plastoquinon (PQ)), changing it from the oxidized form PQ into the reduced and hydrogenated form PQH_2 (Fig. 7.6):

$$\mathrm{PQ} + 2\mathrm{H}^+ \xrightarrow{\hbar\omega,\, 2e^-} \mathrm{PQH_2}$$

This undergoes the following reaction:

Fig. 7.5 Energetic positions of the different molecular subunits related to the NADPH synthesis in photosystem I. Red/Ox indicate the direction of increasing reduction (oxidation) potentials, respectively

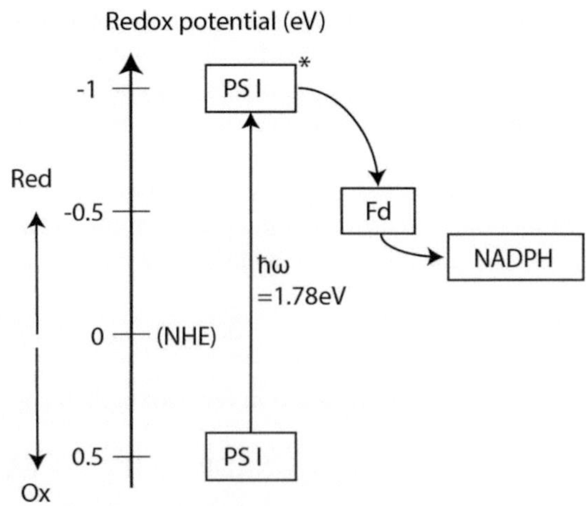

[3] NHE: normal hydrogen electrode $Pt/H_2 : 2H^+ + 2e^- \rightleftharpoons H_2$ at E_{vac} -4,5eV.

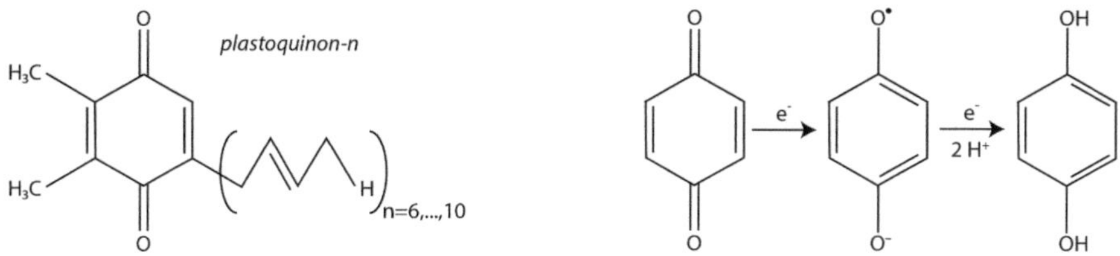

Fig. 7.6 Left: typical structure of the plastoquinon complex. Right: the quinon form of PQ is changed to semiquinon by adding an electron and finally into the quinol form by adding a second electron and two protons

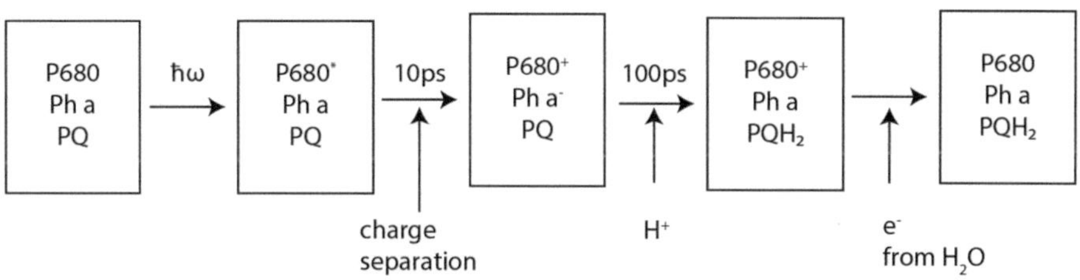

Fig. 7.7 Time scales of the reaction pathway at P680. Following optical excitation, the excited electron is transferred from P680* to Pha in typically 10ps. The subsequent transfer to PQ and the modification to PQH by the reaction with a proton instead takes around 100 ps. Finally, a solvated electron from the water-splitting reaction resets the P680 reaction center back into its ground state

As already mentioned, pheophytin[4] (related to chlorophyll a, but without the Mg center) is responsible for the fast and efficient electron transfer from the excited state of the reaction center $P680^*$ of PS II, occuring spontaneously in typically 10 ps (Fig. 7.7).

The third sub-reaction involves the molecular complex *cytochrome bf (Cyt bf)*. It produces a proton gradient through the thylakoid membrane (see below), powered by the electron flux from PS II to PS I, liberating protons by the oxidation of PQH_2:

$$PQH_2 + 2PCy(Cu^{2+}) \xrightarrow{\searrow 2e^- \nearrow} PQ + 2PCy(Cu^{+}) + 2H^{+}$$

A main functional component of cytochrome bf is the *Plastocyanin Complex (PCy)*, shown in the figure below. It contains a central copper atom, which can change its oxidation state between 2+ and + by exchanging an electron with its environment (Fig. 7.8).

Similar to photosystem I above, Fig. 7.9 below shows the energetic positions of the different components active in photosystem II on a redox potential scale relative to NHE:

The last sub-reaction produces **ATP** via the enzyme ATP-synthase, which catalyses the reaction ADP + P $\rightarrow$ ATP.

[4] Short "Pheo".

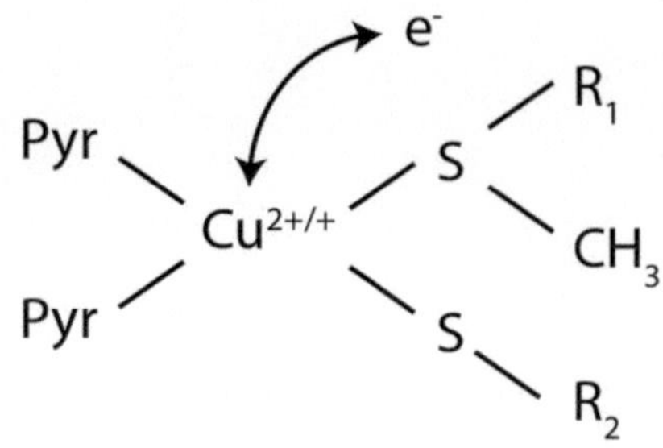

Fig. 7.8 Schematic structure of the plastocyanin complex in cytochrome bf. 'Pyr' symbolizes a pyrrole ring (C_4N)

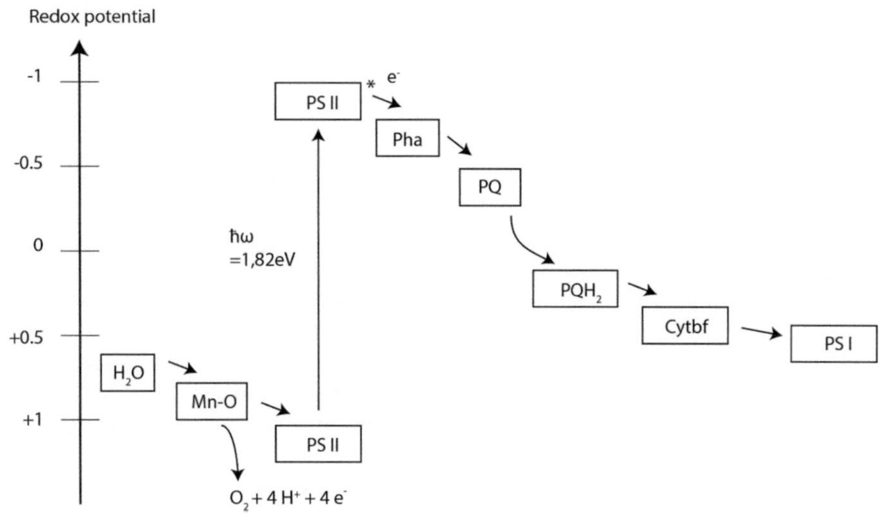

Fig. 7.9 Energy diagram of the different complexes involved in sub-reactions two and three. Mn-O indicates the oxygen-evolving center where water splitting occurs via $P680^+$

The enzyme is powered by a proton flux through it, caused by the proton concentration gradient inside vs. outside of the thylakoid membrane, which is generated by Cyt bf. This leads to a typical pH-value of 4 inside the volume enclosed by the thylakoid membrane (*lumen*) versus a pH-value of 7.5 outside (*stroma*). A summary of all main components active within the thylakoid membrane, the main proton and electron fluxes between these units and the two cyclic reactions involving plastoquinone and plastocyanin are summarized in Fig. 7.10.

To make this very complex photosynthetic machine run optimally, the spatial arrangement of the different trans-membrane proteins is very important. Therefore, the thylakoid membrane consists of (cf. Fig. 7.11):

- Stacked regions containing mainly PS II and Cyt bf. Here, a large part of the generation of the H^+ gradient takes place.

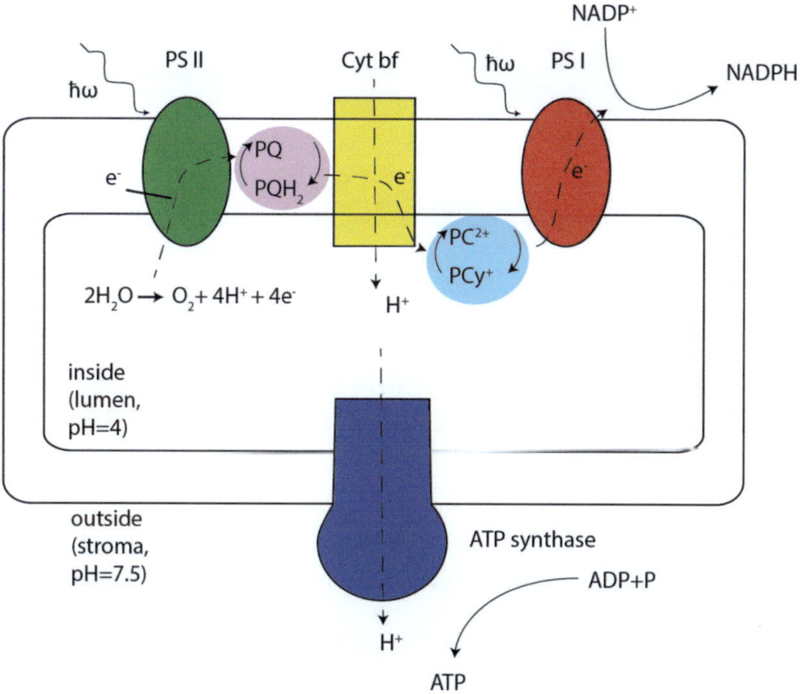

Fig. 7.10 Main functional units of photosynthesis in the thylakoid membrane with the corresponding electron and proton fluxes

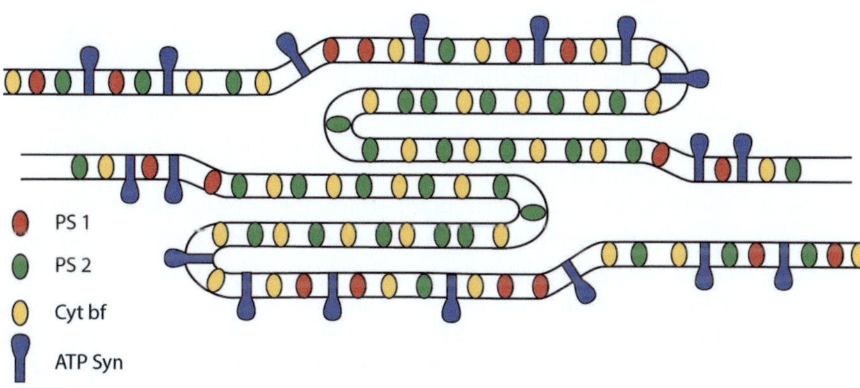

Fig. 7.11 Schematic spatial arrangement of the four main photosynthetic proteins in stacked and non-stacked regions of the thylakoid membrane

- Non-stacked regions with all four proteins, where the generation of NADPH and ATP occurs.

The thylakoid membrane itself is contained in the cell membrane of the chloroplasts in the leaves of plants (Fig. 7.12).

Fig. 7.12 Schematic structure of chloroplast cells with the thylakoid membrane contained inside

7.4 Details of Photon Absorption and Energy Transfer in the Light-Harvesting Complexes of Photosystems

Both photosystems contain an antenna protein (light-harvesting complex (LHC)) consisting of about 50 carotenoid molecules, 200 chlorophyll a and b molecules plus a reaction center (RC). The chlorophyll molecules[5] as the main optical absorption sites are arranged in a torus around the reaction center, as shown schematically in Fig. 7.13.

The optical absorption centers in chlorophyll are four *pyrrole rings* (Pyr, C_4N) surrounding a central Mg atom in a planar structure. Absorption in the red spectral region occurs in the double bonds of these rings as $\pi\pi^*$ (bonding-antibonding) transitions of the ppπ bonds (Fig. 7.14).

The absorption of visible light (red photons) occurs via $\pi - \pi^*$-(bonding-antibonding) transitions of the ppπ bonds (Fig. 7.15):

The exact HOMO[6]-LUMO[7]-splitting depends on the specific ligands bound to the pyrrole rings. The absorption coefficient of chlorophyll is roughly $1 \times 10^5\ \mathrm{cm^{-1}mol^{-1}}$. This means that at full sunlight about 10 photons are absorbed per chlorophyll molecule per second. This excitation frequency is slow enough to provide sufficient time for all subsequent reactions to proceed.

Normally, for these optical electric dipole transitions during the excitation of chlorophyll molecules the electron spin S is preserved: the spin configuration (antiparallel) of the ground state singlet is preserved and leads to a singlet excited state, whereas the total angular momentum $L + S$ is increased by the angular momentum $J = 1$ of the absorbed photon (singlet transitions $S = 0$, $\Delta S = 0$, $\Delta L = 1$). However, as shown in Fig. 7.16, in a few cases one of the electron spins may flip due to spin-orbit coupling (e.g. close to heavy metal atoms), causing a system crossing into the triplet state T* (parallel spins, $S = 1$).

[5] There are many variants of chlorophyll molecules varying by rest groups bound to the pyrrole rings. The discovery of the chlorophyll structure led to three Nobel prizes: Wilstätter 1915, Fischer 1930, Woodward 1965.

[6] Highest Occupied Molecular Orbital.

[7] Lowest Unoccupied Molecular Orbital.

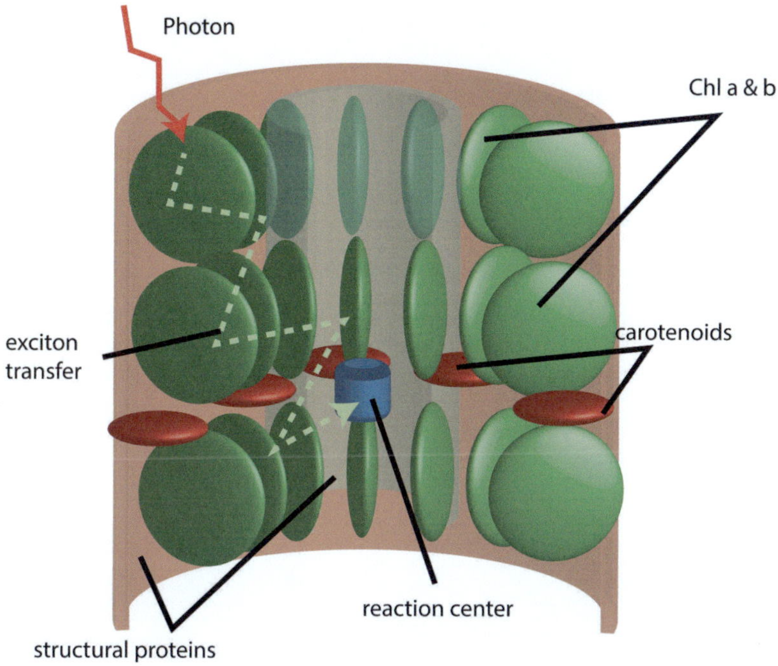

Fig. 7.13 Antenna protein arrangement of the light harvesting complex

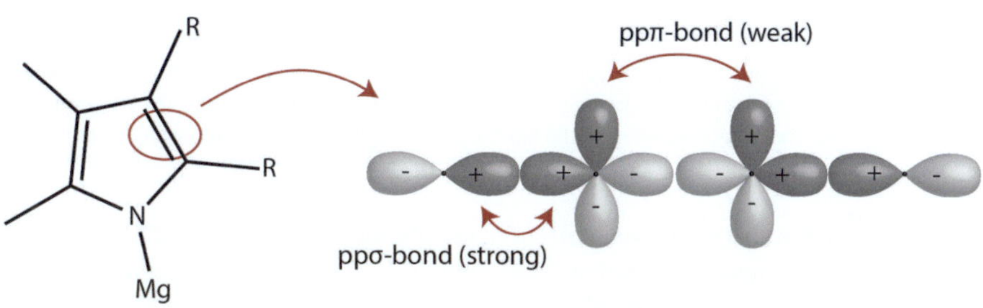

Fig. 7.14 Left: one of the four pyrrole rings (with two C = C double bonds) linked to the central Mg atom in a chlorophyll molecule. Different variants of chlorophyll have different rest groups R attached to the rings. Right: electronic structure of a C = C double bond consisting of carbon p-orbitals arranged in a strongly bound ppσ-chain and a weaker ppπ-bond perpendicular to the chain + and − indicate the opposite phases of the two p-wavefunction lobes

Since molecular oxygen is always present close to the photosystem due to the water splitting reaction, the triplet state of chlorophyll ($S = 1$, $m_s = +1, 0, -1$) can then couple to O_2-molecules by resonant energy and spin exchange (*Dexter transfer*). This causes the formation of very reactive and dangerous singlet oxygen molecules. Singlet oxygen is one of the most reactive oxidation agents known in chemistry and would cause the destruction of the photosystem. To prevent this, the photosystem also contains a number of carotenoid molecules, acting as anti-oxidants to get rid of the dangerous T* or $^1O_2$ states.

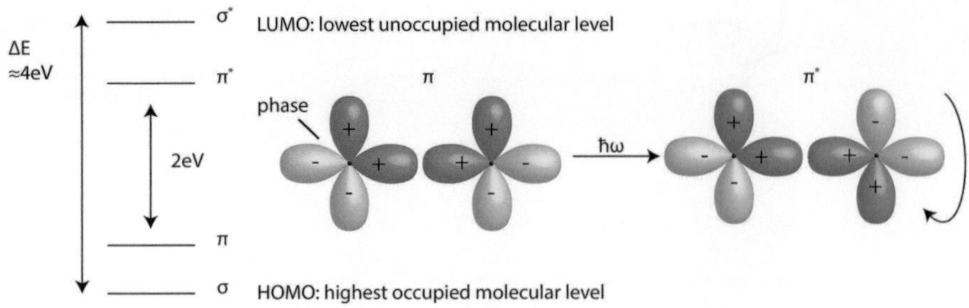

Fig. 7.15 Illustration of a $\pi - \pi^*$-transition in a C = C carbon double bond. The weaker π-bond forms the inner HOMO-LUMO level pair and the optical transition changes the π-orbital configuration from bonding to anti-bonding. + and − indicate the different phases of the p-wave functions

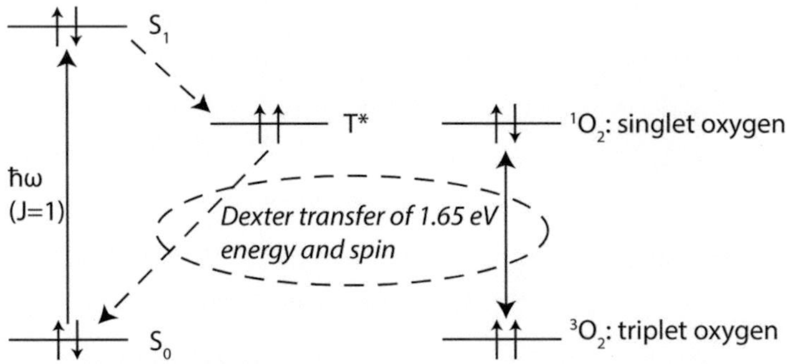

Fig. 7.16 Illustration of a system crossing from the excited singlet stateS_1 into the excited triplet state T* due to spin-orbit coupling. In the presence of O_2 molecules, this enables the occurrence of a resonant Dexter transfer (both energy and spin are exchanged) between chlorophyll and O_2 via the excitation of the oxygen molecule from its triplet ground state to an excited singlet state

To increase the absorption of sunlight in the photosystems, about 200 chlorophyll molecules are arranged in the antenna complexes. In these complexes energy can be transported very rapidly (10ps) via radiationless dipole-dipole interaction between neighbouring chlorophyll molecules. This is called *Förster transfer*. The corresponding transfer rate is related to the spontaneous fluorescence rate of the molecules via:

$$\boxed{k_{transfer} = k_{fluorescence} \left(\frac{R_0}{r}\right)^6} \tag{7.4.1}$$

Here r is the distance between neighboring molecules and $R_0^6 = 8.8 \times 10^{-5} \mathrm{J}\kappa^2 n^{-4}$ (n: index of refraction). J is the so-called *energy-overlap factor*, κ the orientation factor:

$$J \equiv \int \alpha(\lambda) F_D(\lambda) \lambda^4 \mathrm{d}\lambda$$

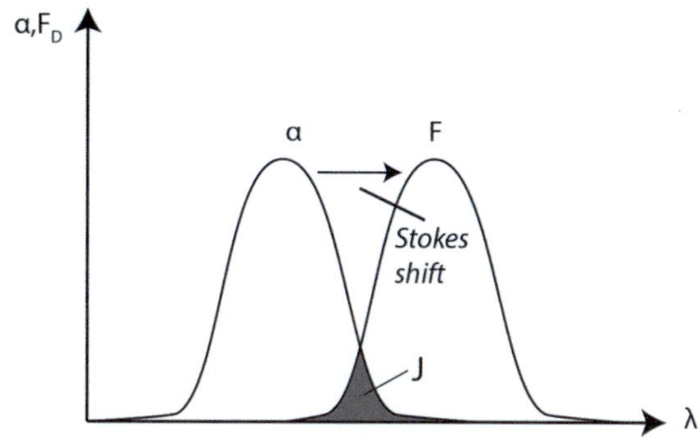

Fig. 7.17 Schematic illustration of the typical Stokes-shift between absorption and emission spectra in molecules. The shaded area contributes to the energy-overlap factor J

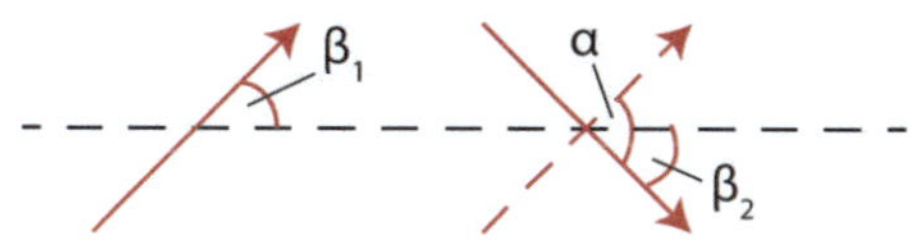

$$\kappa = \cos(\alpha) - 3\cos(\beta_1)\cos(\beta_2)$$

Fig. 7.18 Orientation factor κ, determined by the relative angles of the two interacting dipole moments

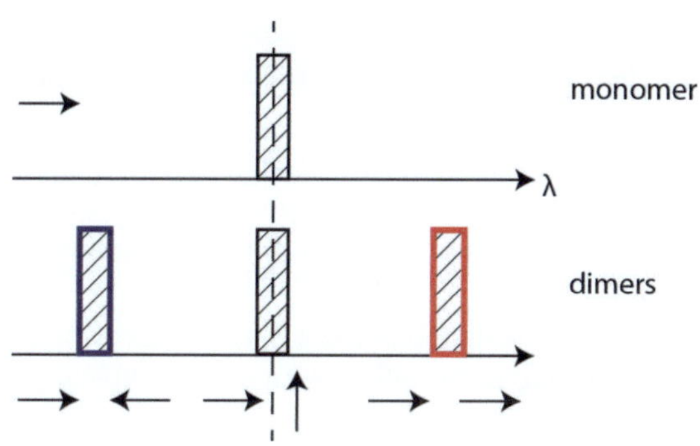

Fig. 7.19 Orientation-dependent spectral shift of molecular dimers coupled by dipolar interaction. Compared to the monomers, a red shift occurs for parallel and a blue shift for anti-parallel dipole orientations

where $\alpha(\lambda)$ is the absorption coefficient spectrum of the receiving molecule, while $F_D(\lambda)$ is the fluorescence emission spectrum of the emitting molecule. Also, the relative orientation of interacting dipoles influences the value of R_0 via the orientation factor (Figs. 7.17, 7.18).

Typical values for R_0 in antenna complexes are 6–10 nm. The relative orientation of the dipole moments also leads to an energy shift of the interacting molecules (Fig. 7.19):

The efficient Förster transfer between the chlorophyll molecules in an antenna complex allows all molecular excitations to finally be collected by a central reaction center with a lower HOMO-LUMO-splitting (Fig. 7.20).

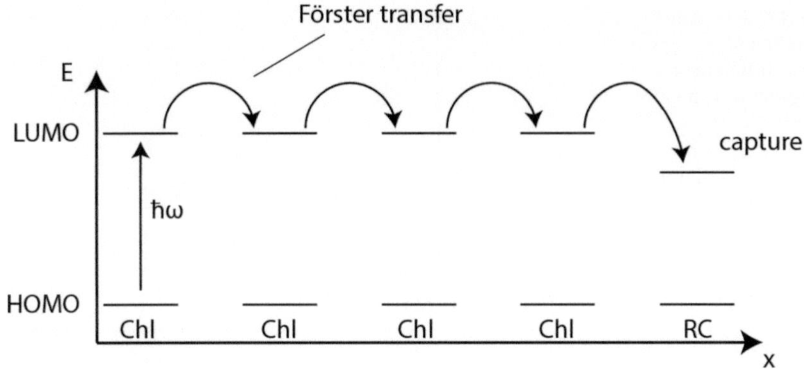

Fig. 7.20 Fast Förster transfer between chlorophyll molecules in an antenna complex and final capture of the excitation in a reaction center (RC) with smaller HOMO-LUMO splitting

7.5 Technical Use of Biomass

Biomass usually comes from different sources in different forms. Therefore, the typical energy content of the starting material (dry mass) can vary a lot (Table 7.2):

There are also different forms in which the energy can be extracted from biomass:

- physically: direct combustion in ovens or heat engines
- chemically: gasification, pyrolysis, carbonisation
- biologically: fermentation, fowling, feedstock for animals

A detailed discussion of the use of biomass would go beyond the scope of this book. Therefore we just briefly mention some typical examples in the box below:

Table 7.2 Energy content in GJ per ton of the dry mass of various biomass sources

Component	Energy content	Resources
Starch, sugar	$\approx 15\,\mathrm{GJt^{-1}}$	Corn, sugar cane, sugar beets
Cellulose	$\approx 5 - 15\,\mathrm{GJt^{-1}}$	Grass, wood, ...
Plant oils	$\approx 40\,\mathrm{GJt^{-1}}$	Soy, sun flowers, palms
Algae	$\approx 35\,\mathrm{GJt^{-1}}$	

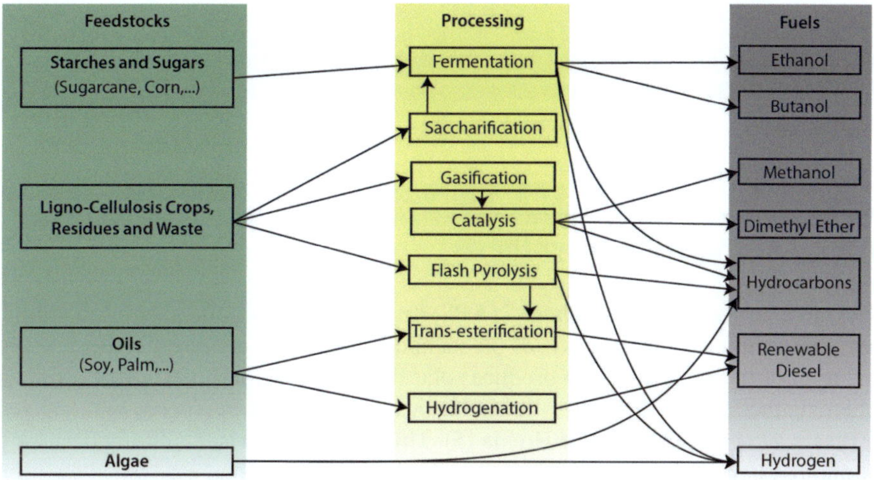

Fig. 7.21 Overview of fuel production by different processing methods using different biomass feedstocks

Typical uses of biomass:

- Direct combustion: here the efficiency of extraction of energy varies widely between 5% (open fireplaces) and 50% (bio-power-plant with waste heat usage)
- Gasification: this is the reaction of biomass with water steam and air/O_2 at temperatures between 300–1000 °C at pressure of 1–30 bar. The main product is *syngas*, a mixture of C_O and H_2, from which most carbohydrates can be synthesised.
- Pyrolysis: the processing of biomass without air/C_O at elevated temperatures of 300–500 °C until all volatile components have disappeared. The main product is charcoal which can be burnt to CO_2 to almost 100%. The energy density of charcoal is twice that of the starting material.
- Bio-ethanol by fermentation: this produces normal alcohol C_2H_5OH, made out of sugar or starch, followed by distillation. The achievable output is about 200L per ton biomass.

  The following Table gives an impression of current volumes of bio-ethanol produced in different regions:

Country	Production volume
USA	$40 \cdot 10^9 L$
Brazil	$30 \cdot 10^9 L$
EU	$5 \cdot 10^9 L$
GER	$1 \cdot 10^9 L$

This amounts to about 1% of the worldwide oil production of 6000 GL (Fig. 7.21).

7.6 Artificial Photosynthesis

The term "artificial photosynthesis" (APS) describes the scientific attempts to mimic the photosynthesis mechanisms of plants and bacteria in a technical context, using the light of the sun directly to produce solar fuels. The simplest version of artificial photosynthesis is solar water splitting for the production of hydrogen, first realized 50 years ago by Fujishima and Honda [5]. This mimics the water splitting occurring at photosystem II described in Sect. 7.2. As the initial step, one uses the absorption of sun light in a semiconductor or dye molecules to excite electron-hole pairs. The excited holes are then used either directly or with the help of suitable catalysts for water oxidation:

$$2H_2O + 4h^+ \longrightarrow O_2 + 4H^+$$

This is shown on the left side of Fig. 7.22 and constitutes an anodic oxidation reaction consuming the holes in the anode. The remaining excited electrons are transported via an electrical connection to the nearby cathode, where they can be used to reduce the protons generated at the anode to hydrogen:

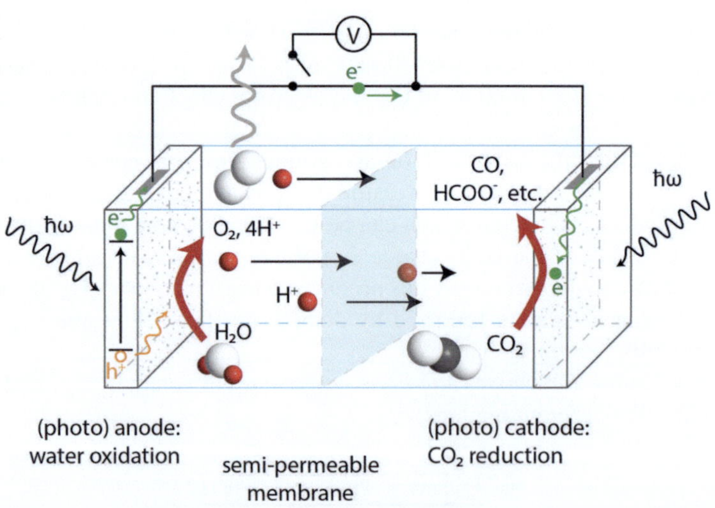

Fig. 7.22 Electrode arrangement for water splitting and CO_2 reduction for artificial photosynthesis using solar irradiation. See text for details. Alternatively, an external electrical power source can be used to drive the corresponding reactions electrochemically

$$4H^+ + 4e^- \longrightarrow 2H_2$$

This completes the water splitting reaction. A semi-permeable membrane between the electrodes can be used to spatially separate the evolution of O_2 and H_2. Using optimized photo-anodes with the right band gap and position of the hole energy level, efficiencies up to 14% have been achieved under realistic solar irradiation conditions so far.

A similar electrode set-up is also used in modern electrolyzers, where an external voltage U drives the electron current, using the electrical power e.g. generated by wind mills or solar cells for water electrolysis. In this case, the anode does not need to be photo-sensitive and can just be a metal plate like the cathode. Such electrolyzers have reached efficiencies for electrical water splitting around 70%. Combined with a solar cell of 20% efficiency, this also gives an efficiency of 14% for overall water splitting, with practical advantages such as better long-term stability and larger operational flexibility for the combined solar cell/electrolyzer approach.

This is one of the reasons why cathodic reactions other than proton reduction have gained significant interest for artificial photosynthesis in the last years. One prominent example is the combination of anodic water oxidation with cathodic reduction of CO_2, constituting a simplified version of the Calvin cycle in Fig. 7.2. For example, CO_2 can be reduced to formic acid by the reaction:

$$CO_2 + H^+ + 2e^- \longrightarrow HCOO^-$$

The formic acid then can be further processed to other molecules for long term energy storage. Another possibility is the reduction of CO_2 to CO:

$$2CO_2 + 4H^+ + 4e^- \longrightarrow 2CO + 2H_2O$$

Together with solar hydrogen, this allows the production of syngas ($H_2 + CO$) as a starting point for other solar fuels such as methanol:

$$CO + 2H_2 \longrightarrow CH_3OH$$

Further Reading

- CO_2-levels monitored by NOAA's Global Monitoring Lab: www.climate.gov
- Data on worldwide use of bioenergy: www.iea.org/data-and-statistics/
- Hou, H.J.M., Najafpour, M.M., Moore, G.F., & Allakhverdiev, S.I. (eds.): Photosynthesis: Structures, Mechanisms, and Applications. Springer International Publishing (2017).
- Kan, T., Strezov V., Evans, T.J.: Lignocellulosic biomass pyrolysis: a review of product properties and effects of pyrolysis parameters. Renew. Sust. Energ. Rev. **57**, 1126–1140 (2016).

- Sansaniwal, S.K., et al.: Recent advances in the development of biomass gasification technology: a comprehensive review. Renew. Sust. Energ. Rev. **72**, 363–384 (2017).
- Balat, M., Balat, H., Öz, C.: Progress in Bioethanol Processing. Prog. Energy Combust. Sci. **34**, 551–573 (2008).
- Zhang, B., Sun, L.: Artificial photosynthesis: opportunities and challenges of molecular catalysts. Chem. Soc. Rev. **48**, 2216–2264 (2019).

References

1. Green, C., Byrne, K. A.: Biomass: Impact on Carbon Cycle and Greenhouse Gas Emissions. In: C. J. B. T.-E. of E. Cleveland (ed.), pp. 223–236 (2004)
2. BP.: Statistical review of world energy 2021 **70** (2021)
3. IEA.: World Energy Balances. https://www.iea.org/data-and-statistics/. Accessed 30 July 2022
4. Energieatlas Bayern, Bayerisches Staatsministerium für Wirtschaft, Landesentwicklung und Energie, https://www.energieatlas.bayern.de/thema_biomasse/daten.html. Accessed 30.07.2022
5. Fujishima, A., Honda, K.: Electrochemical photolysis of water at a semiconductor electrode. Nature **238**, 37–38 (1972)

8 Photovoltaics

Abstract

This chapter deals with the physics of solar cells. The basic photonic processes such as adsorption, reflection, emission and the charge carrier processes separation, thermalisation, recombination and extraction are explained with the corresponding theory and examples. The next subchapter presents the various types of solar cells and their differences in generating electricity. Later, the current-voltage characteristics are explained in a view to the layout of solar cells and their maximum efficiency. The chapter concludes with the main loss mechanisms in solar cells and the different routes to improve efficiency.

8.1 General Considerations

Photovoltaics is the direct transformation of light into electricity. It is a purely optoelectronic process in semiconductors without any chemical reactions, mechanical movements or use of heat involved. Together with wind power, photovoltaics is considered the major form of renewable energy production for a sustainable future. Accordingly, the worldwide installed photovoltaic power has increased from about $30\,GW_p$[1] in 2010 to more than $840\,GW_p$ in 2021, with currently more than $100\,W_p$ additional power capacity newly installed every year [1]. At the same time, due to mass production, the cost per k W_p installed photovoltaic power has decreased to about 1000€ in 2020, enabling electricity production at a cost of currently about 4 ct/kWh.

As already discussed in Chap. 6, the amount of harvestable solar energy depends strongly on the specific geographic location. As shown in Fig. 8.1, already in Europe the available average annual solar energy density can vary by a factor of three between Scandinavia and Mediterranean regions.

[1] Here W_p means Watt peak, i.e. the maximum power generated under ideal standard illumination conditions: AM 1.5, $1000\,W\,m^{-2}$ incident irradiation intensity, cell temperature $25\,°C$.

M. Stutzmann and C. Csoklich, *The Physics of Renewable Energy*,
Graduate Texts in Physics,
https://doi.org/10.1007/978-3-031-17724-8_8

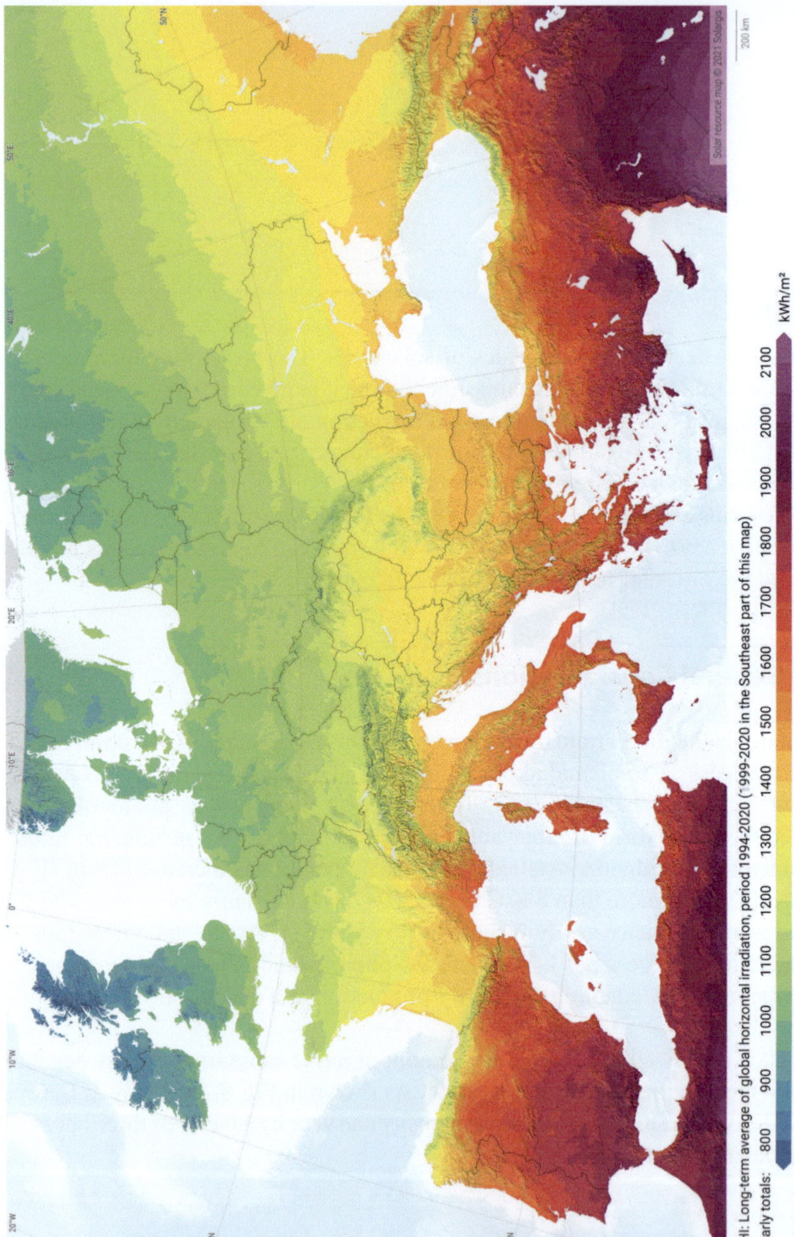

Fig. 8.1 Average annual sum of solar power (in [kW h m^{-2}]) available in Europe and Northern Africa. Solargis is gratefully acknowledged for the permission for reproduction (©2020 The World Bank, *Source* Global Solar Atlas 2.0, Solar resource data: Solargis)

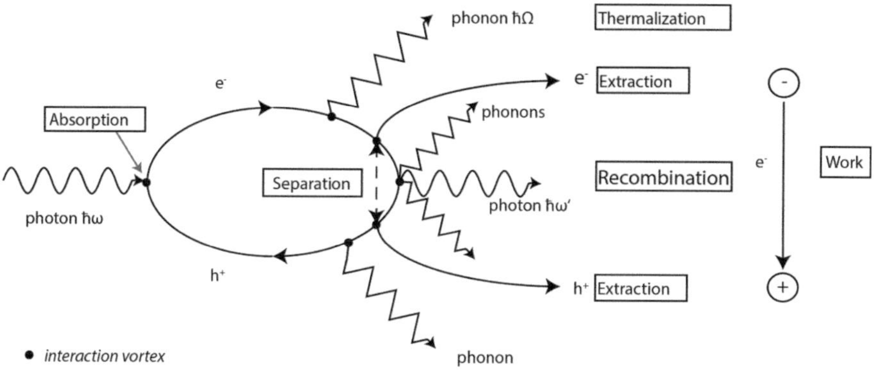

Fig. 8.2 Feynman diagram of photovoltaic energy conversion. Dots indicate interaction vortices where the indicated physical processes need to be optimized

To make photovoltaics profitable also in regions with solar energy densities around 1000 kW h m^{-2}, high conversion efficiencies of solar cells on an industrial scale are necessary. The complexity of this task can be understood by looking at the Feynman diagram of photovoltaic energy conversion in Fig. 8.2.

The optimisation of photovoltaic devices requires the optimisation of many materials and device parameters. Absorption of solar photons competes with reflection and transmission, which have to be minimized. The energy stored in photoexcited electrons and holes is limited by thermalization losses (emission of phonons) depending on the mismatch between photon energies and the band gap of the chosen semiconductor absorber layers. This stored energy can be lost again by recombination of the excited carriers via emission of phonons and photons. This has to be prevented by an efficient spatial separation of electrons and holes, similar to the first step in photosynthesis. Finally, electrons and holes have to be extracted via suitable contacts, minimizing electrical losses in a solar cell. In the following, we will discuss challenges and optimization strategies for efficient solar cells in more detail.

8.2 Basic Processes in Photovoltaics

8.2.1 Photons

Photons are characterized by two properties:

energy $hf = \hbar\omega$
where f and ω are the frequency and angular frequency.

momentum $\mathbf{p} = \hbar\mathbf{k}$
with $\mathbf{k}$ the wavevector ($\mathbf{k} = \frac{2\pi}{\lambda} = \frac{\omega}{c} = \frac{2\pi f}{c}$ ($\Rightarrow \lambda = c/f$) and c, the velocity of light.

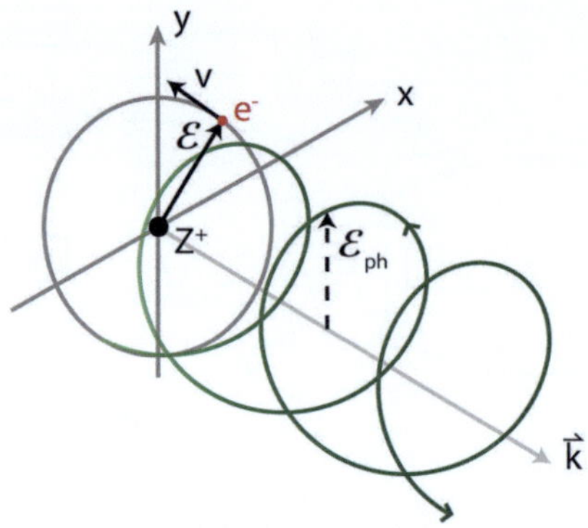

Fig. 8.3 Circular polarisation of the EM-wave corresponding to a photon. Classically, the circular polarization can be related to the circulating electrical field $\mathcal{E}$ between an electron on its orbit around the positively charged nucleus. Non-classically, the orbital momentum of the circling electron is quantized in terms of $\hbar$

In addition, photons are bosons with angular momentum $J = 1$, realized by circular polarization of the corresponding electromagnetic wave (Fig. 8.3).

This leads to three polarisation states:

$$J = 1 \begin{cases} |m_j = +1\rangle\ (\uparrow) & \text{right circular polarisation} \\ |m_j = 0\rangle\ \left(\frac{1}{\sqrt{2}}((\uparrow) + (\downarrow))\right) & \text{linear polarisation} \\ |m_j = -1\rangle\ (\downarrow) & \text{left circular polarisation} \end{cases} \tag{8.2.1}$$

Summary of basic properties of photons as planar electromagnetic waves:

$$\mathcal{E} = \mathcal{E}_0 \exp\left[i(\mathbf{kr} - \omega t)\right]; \qquad \mathcal{B} = \mathcal{B}_0 \exp\left[i(\mathbf{kr} - \omega t)\right]$$

$$\mathcal{E}_0 \perp \mathcal{B}_0; \qquad \mathcal{E}_0, \mathcal{B}_0 \perp \mathbf{k}; \qquad \mathcal{E} \text{ and } \mathcal{B} \text{ in phase}; \qquad \mathcal{E}_0 = c\mathcal{B}_0$$

$$c_0 = \frac{1}{\sqrt{\varepsilon_0 \mu_0}}; \qquad c = \frac{1}{\sqrt{\varepsilon \varepsilon_0 \mu \mu_0}} = \frac{c_0}{n} \text{ in matter } (n = \sqrt{\varepsilon} \text{ for } \mu = 1)$$

with n the refractive index.

8.2.2 Photon Density of States (DOS)

Photons confined to a volume $V = L_x L_y L_z$ have a quantized momentum $\mathbf{p}$:

$$\mathbf{p} = (p_x, p_y, p_z) = \left(\frac{i\hbar}{L_x}, \frac{j\hbar}{L_y}, \frac{k\hbar}{L_z}\right); \qquad i, j, k = 0, \pm 1, \pm 2, \ldots$$

This follows from the Heisenberg uncertainty principle for position and momentum:

$$\Delta x \Delta p_x \geq h \qquad \text{analogous for } y \text{ and } z \tag{8.2.2}$$

Therefore, the minimal phase space volume of one quantized state is:

$$\Delta x \Delta y \Delta z \Delta p_x \Delta p_y \Delta p_z \geq h^3$$

Due to the linear photon dispersion relation $\hbar\omega = cp$, this leads to the following relations:

- The number $N_\gamma(\hbar\omega)$ of photons with energy $\leq \hbar\omega$ (with a factor of two originating from the polarisation) is given by:

$$N_\gamma(\hbar\omega) = 2V \frac{\frac{4\pi}{3}|\mathbf{p}|^3}{h^3} = \frac{8\pi}{3} V \frac{(\hbar\omega)^3}{(hc)^3} \tag{8.2.3}$$

- This results in a photon density of states (DOS):

$$D_\gamma(\hbar\omega) \equiv \frac{1}{V}\frac{\mathrm{d}N_\gamma(\hbar\omega)}{\mathrm{d}\hbar\omega} = \frac{(\hbar\omega)^2}{\pi^2\hbar^3c^3} \tag{8.2.4}$$

- Therefore, in thermal equilibrium the spectral distribution of photons (i.e. the spectral density $\mathrm{d}n_\gamma(\hbar\omega)$ of photons with energy between $\hbar\omega$ and $\hbar\omega + \mathrm{d}\hbar\omega$) is:

$$dn_\gamma(\hbar\omega) = D_\gamma(\hbar\omega) f_\gamma(\hbar\omega) d\hbar\omega \tag{8.2.5}$$

  with

$$f_\gamma(\hbar\omega) = \left[\exp\left(\frac{\hbar\omega}{kT}\right) - 1\right]^{-1} \tag{8.2.6}$$

  where (8.2.6) is the equilibrium occupation function for bosons.
  For an isotropic distribution of propagation directions (e.g. photons in a hollow metal sphere as a black body), the density of states into the solid angle $\mathrm{d}\Omega$ is:

$$D_{\gamma,\Omega}(\hbar\omega) = \frac{1}{4\pi} D_\gamma(\hbar\omega) \tag{8.2.7}$$

  This finally gives for the photon energy per unit volume and energy interval $\mathrm{d}\hbar\omega$ into the solid angle $\mathrm{d}\Omega$ ($\equiv$*photon spectrum*):

$$\mathrm{d}e_\gamma(\hbar\omega) = D_\gamma(\hbar\omega) f_\gamma(\hbar\omega)\hbar\omega\, \mathrm{d}\hbar\omega\, \mathrm{d}\Omega \tag{8.2.8}$$

$$= \frac{(\hbar\omega)^3}{4\pi^3\hbar^3(c_0/n)^3}\left[\exp\left(\frac{\hbar\omega}{kT}\right) - 1\right]^{-1} \mathrm{d}\hbar\omega\, \mathrm{d}\Omega \tag{8.2.9}$$

Note that the photon density of states in matter is proportional to n^3. When going from a medium n_1 to a medium with $n_2 \neq n_1$ and keeping the distribution function $f_\gamma(\hbar\omega)$ constant, this requires that at the interface a part of the photons has to be reflected

8.2.3 Absorption, Reflection, Emission

Solar cells are generally built with diamagnetic[2] semiconductors as the absorbers. This means that the interaction with photons mainly occurs via their electric field $\mathcal{E}$. The principal process is acceleration of electrons by the electric field. Thereby an electron will take up an energy $E_e = |\mathcal{E}|e \cdot x$, where x is the spatial length of undisturbed acceleration before a scattering event takes place. Thus, the interaction strength is given by the dipole moment $e \cdot x$ of electrons in the absorbing medium. Similarly, the quantum mechanical matrix element for optical absorption contains the dipole operator, and the transition probability W for transitions from the initial ground state $|i\rangle$ to the excited state $|f\rangle$ is proportional to the light intensity I:

$$W \propto |\mathcal{E}\langle i|e \cdot x|f\rangle|^2 = \mathcal{E}^2 e^2 |\langle i|x|f\rangle|^2 \propto I$$

In general, the intensity I of an electromagnetic wave is given by:

$$I = \frac{1}{\mu_0}|\mathcal{E} \times \mathcal{B}| = \frac{1}{\mu_0}\mathcal{E}_0\mathcal{B}_0 = \frac{1}{\mu_0 c}\mathcal{E}_0^2 = \sqrt{\frac{\varepsilon_0}{\mu_0}}\mathcal{E}_0^2 \tag{8.2.10}$$

or, in the photon particle model

$$I = \hbar\omega\Phi \tag{8.2.11}$$

with the photon flux density (number of photons per area and time) Φ.

Absorption of a photon is subject to three selection rules:

1. energy conservation: $E_f - E_i = \hbar\omega$
2. momentum conservation: $\mathbf{p}_f - \mathbf{p}_i = \hbar\mathbf{k} \approx 0$ (due to the very small momentum of massless photons compared to electrons)
3. angular momentum conservation: $L_f - L_i = \hbar$

[2] Relative magnetic permeability $\mu = 1$.

- In atoms, angular momentum conservation leads to the dipole selection rule: $\Delta n = \pm 1$ for the main quantum number n.
- In solids (esp. semiconductors) angular momentum is conserved preferably by generation of *excitons* (bound electron-hole pairs).

The optical properties of solids are parametrized by the complex dielectric function $\tilde{\varepsilon}(\hbar\omega)$ or the complex index of refraction $\tilde{n} = \sqrt{\tilde{\varepsilon}}$. Traditionally, these are written in the form:

$$\tilde{\varepsilon} = \varepsilon_1 + i\varepsilon_2 \qquad \text{and} \qquad \tilde{n} = n + i\kappa \tag{8.2.12}$$

where n is the index of refraction and κ the absorption constant. For electromagnetic waves in matter the complex index of refraction then leads to the following changes compared to vacuum (k = wavenumber):

$$c_0 \rightarrow \frac{c_0}{\tilde{n}} \Rightarrow k = \frac{\omega}{c_0} \rightarrow k = \frac{\omega}{c_0}\tilde{n}$$

This gives for the electric field of a plane wave:

$$\mathcal{E} = \mathcal{E}_0 \exp\left[i\left(\frac{\omega}{c_0}(n + i\kappa)x - \omega t\right)\right] \tag{8.2.13}$$

$$= \mathcal{E}_0 \exp\left[-\frac{\omega}{c_0}\kappa x\right] \exp\left[i\frac{\omega}{c_0}nx - \omega t\right] \tag{8.2.14}$$

so that the intensity $I \propto |\mathcal{E}|^2$ follows the law of *Lambert–Beer*:

$$I(x) = I_0 \exp[-2\frac{\omega}{c_0}\kappa x] \equiv I_0 \exp[-\alpha x] \tag{8.2.15}$$

with $\alpha \equiv 2\frac{\omega}{c_0}\kappa$ the *absorption coefficient.*

Note that according to this law 95% of the incoming intensity are absorbed within a distance of $x = 3/\alpha$.

Typical values for the 95% absorption depth of photons with an energy of $\hbar\omega = 2\,\text{eV}$ (red-yellow) in some semiconductor absorber materials commonly used for solar cells are listed in Table 8.1.

The relevant absorption coefficients depend strongly on the nature of possible optical transitions close to the optical band gap of the different absorber materials. As shown in Fig. 8.4, one distinguishes between direct and indirect band gaps for

Table 8.1 95% absorption depths of photons with $\hbar\omega = 2\,\text{eV}$ for various semiconductor absorber materials

Material	α (2 eV) [cm^{-1}]	x (95%) [μm]
GaAs	$3 \cdot 10^4$	1
Crystalline Si	3×10^3	10
Amorphous Si	2×10^4	1.5
Perovskite (PbI) and organic (PCBM/P3HT)	5×10^4	0.6

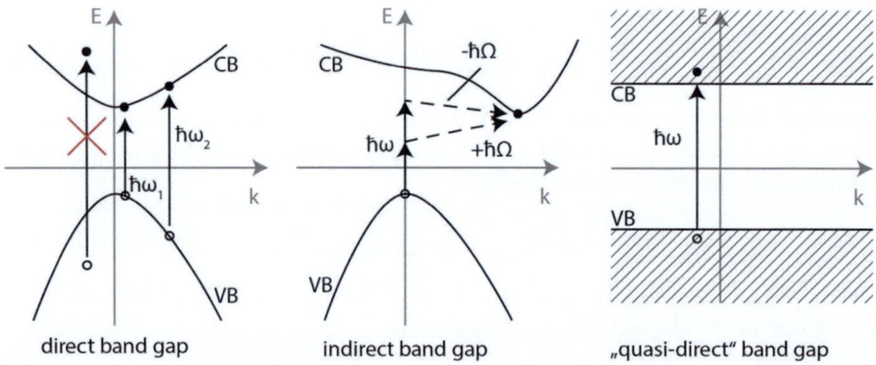

Fig. 8.4 Possible optical transitions in the $E(k)$ band diagrams close to the optical band gap of direct and indirect crystalline semiconductors as well as disordered "quasi direct" semiconductors. CB and VB denote the lowest conduction and the highest valence band, respectively. ω stands for a photon transition, Ω for a phonon transition. See text for further details

crystalline semiconductors and "quasi-direct" band gaps in disordered (amorphous) semiconductors.

In a direct semiconductor, the maximum of the upper valence band and the minimum of the lowest conduction band occur at the same wavevector, typically at $k = 0$ in the center of the Brillouin zone. Important examples of direct semiconductors are GaAlAs, GaInN, ZnO and perovskites. In these materials, direct optical transitions can occur between discrete states of the valence and conduction band (VB, CB) with the same k-value, as indicated by black arrows in Fig. 8.4. Note that compared to electron and hole wavevectors, the wavevector of the massless photons is negligible, so that optical transitions occur vertically up in the $E(k)$ band diagram. Also the electronic states only exist within the discrete bands, so that transitions such as the one indicated by the left-most arrow in Fig. 8.4 are not allowed.

In indirect semiconductors, the two band extrema occur at different k-values, so that close to the band gap no direct optical transitions are possible. Instead, for k-conservation, at least one phonon needs to be involved in the transition to provide the difference in momentum between the initial electron at the top of the valence band and the excited electron at the bottom of the conduction band. The phonon participation can either occur as phonon emission starting from a higher photon energy or as phonon absorption starting from a lower photon energy. Whereas phonon emission can occur at all temperatures, phonon absorption only contributes at higher temperatures, when sufficient phonons are thermally excited. In either case the need of additional phonons lowers the probability of optical transitions, so that the absorption coefficients of indirect semiconductors close to the band gap are generally smaller than those of direct semiconductors. Important examples of indirect semiconductors are Si, Ge, and SiC.

In non-crystalline, disordered semiconductors such as amorphous silicon or most organic semiconductors, the wavevector k is no longer a valid quantum number due to the lacking crystalline periodicity of the atom positions. Instead, the electronic bands $E(k)$ are projected onto the energy axis, giving rise to the electronic density of states $D(E)$, indicated by the hashed regions in Fig. 8.4. Since k-conservation is no longer required, optical transitions occur again vertically up in energy. This is why these transitions are called "quasi-direct".

Reflection of incident photons at the absorber surface occurs due to the change of the complex index of refraction, $\tilde{n} = n + ik$. In the following, we summarize the typical behavior of the reflectivity R, the absorbance A and the transmission T for the case of a solar cell with normal incidence of light from air ($n_1 = 1, \kappa_1 = 0$) into an absorber (n_2, κ_2) without multiple internal reflection due to a sufficiently strong absorption in the absorber layer with thickness d. We then have:

$$R = \frac{(n_2 - 1)^2 + \kappa_2^2}{(n_2 + 1)^2 + \kappa_2^2} \tag{8.2.16}$$

$$A = (1 - R)\,(1 - \exp[-\alpha d]) \tag{8.2.17}$$

$$T = (1 - R)\exp[-\alpha d] \tag{8.2.18}$$

Thus, an incoming photon flux Φ_0 generates a volume density of excited states per unit time of:

$$\boxed{G = (1 - R)\frac{1}{d}(1 - \exp[-\alpha d])\Phi_0} \tag{8.2.19}$$

This is called the average *generation rate* with units cm^{-3} s^{-1}. Whereas the absorber thickness d has to be optimized with respect to the absorption coefficient α, the negative influence of the reflectivity R on the generation rate G has to be minimized by dedicated anti-reflection layers on top of the absorber. To this end, different strategies can be used, as shown for the specific example of crystalline Si solar cells in Figs. 8.5, 8.6 and 8.7.

1. Macroscopic surface texturing with feature sizes larger than the wavelength of light by anisotropic etching (e.g. for c-Si, ZnO, cf. Fig. 8.5)
2. Nano-texturing with feature sizes $<< \lambda$ (cf. Fig. 8.6)
3. $\lambda/4$-anti-reflection coatings (e.g. Si_3N_4 on Si, cf. Fig. 8.7)

Some experimental results for Si wafers textured in these different ways to suppress their reflectivity are shown in Fig. 8.8.

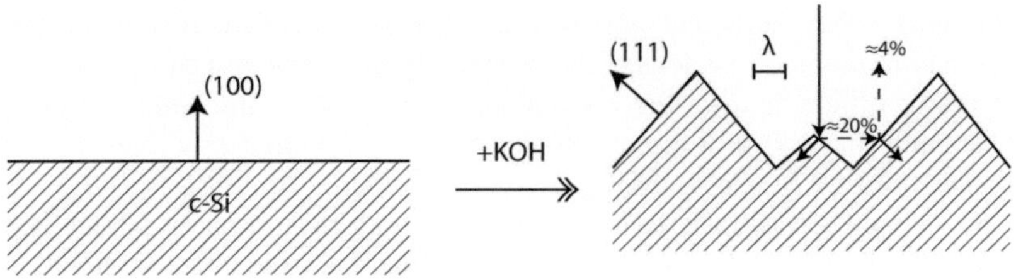

Fig. 8.5 Macroscopic anisotropic etching of a smooth Si surface. In this specific example, a crystalline Si (100) surface is etched by KOH, which preferentially leads to exposure of pyramids with (111) orientation of the crystal surface. The wavelength λ of incoming light is much smaller than the obtained feature sizes, so that the anti-reflection properties result from multiple non-normal reflections within the surface texture

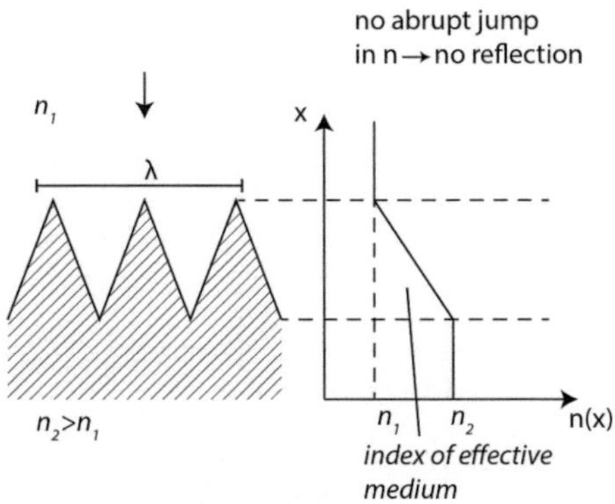

Fig. 8.6 Nano-textured surfaces with feature sizes much smaller than the optical wavelength λ. The resulting intermediate layer forms an effective medium exhibiting a gradual transition from the outside (index n_1) to the inside (n_2) with no jump of the refractive index, thus avoiding reflection

The thermal **emission** of an absorber is obtained by integrating over the photon energy density $\mathrm{d}e_\gamma(\hbar\omega)$ in a given volume (cf. (8.2.8)) and substituting the integrand:

$$x = \frac{\hbar\omega}{k_B T}$$

where T is the absolute temperature and k_B the Boltzmann constant. This results in:

$$e_\gamma = \frac{(k_B T)^4}{4\pi^3\hbar^3\left(\frac{c_0}{n}\right)^3} \underbrace{\int_0^\infty \frac{x^3}{e^x - 1}\mathrm{d}x}_{\pi^4/15} \cdot 4\pi = \frac{\pi^2 k_B^4}{15\hbar^3\left(\frac{c_0}{n}\right)^3} T^4 \tag{8.2.20}$$

Inserting this into the energy flux density $j_e = 1/4\, c e_\gamma$ yields the Stefan-Boltzmann-law for the emission of a black body already used in Chaps. 3 and 6. In addition, optically excited semiconductors emit so-called *luminescence* by radiative recombination of excited carriers in the form of characteristic luminescence spectra (cf. Fig. 8.12).

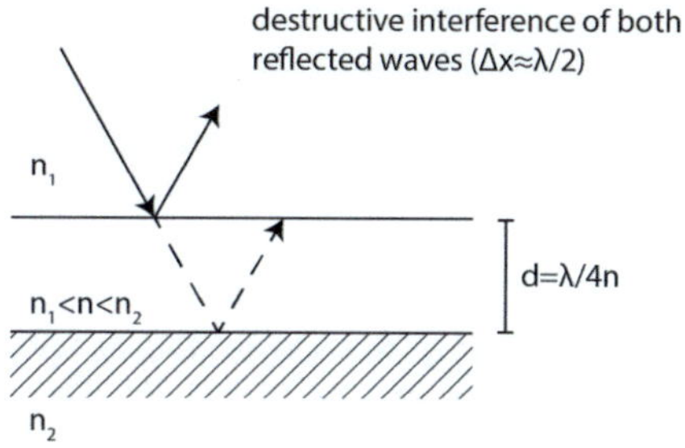

Fig. 8.7 $\lambda/4$-anti-reflection coating formed by a thin film with refractive index n and optical thickness $d = \lambda/4n$ between the outer layer (n_1) and the inner layer (n_2). Due to their optical path length difference $\Delta x = \lambda/2$, the two waves reflected from the two interfaces interfere destructively, thus canceling the reflectivity for the specific wavelength λ

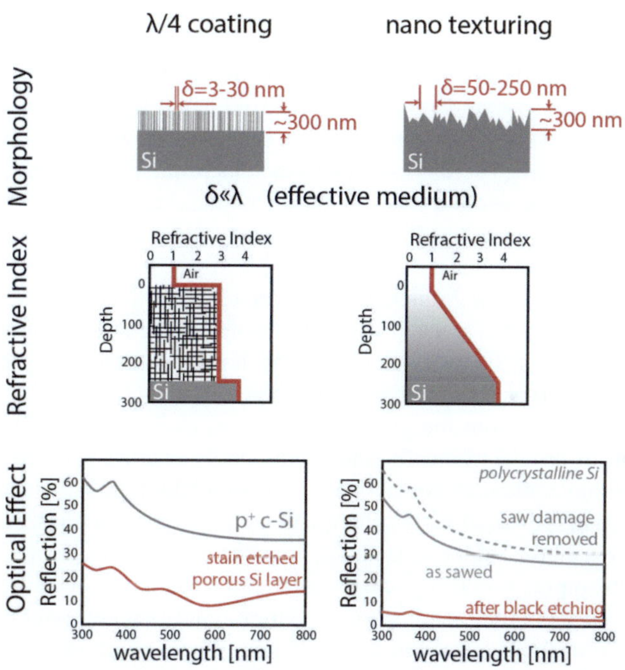

Fig. 8.8 Anti-reflection coatings with different morphologies (upper row, feature size $\delta << \lambda$), resulting in different profiles of the refractive index (middle row). Experimental results for the reflectivity in the wavelength range between 300 and 800 nm (lower row). Left column: creation of a $\lambda/4$ layer by partial removal of the silicon via nanopore etching. Right column: nano-texturing of a surface layer with smooth transition of the refractive index ("black etching") [6]

8.2.4 Thermalization

After absorption of a photon with an energy above the band gap (for semiconductors) or the HOMO-LUMO gap (for molecules), the absorbing material is in a short-lived highly excited state, from which it relaxes very quickly ($\approx 10^{-12}$ s) to a state of zero kinetic energy, with just the band gap energy remaining as potential energy. The excess kinetic energy is lost in the form of molecular vibrational energy or lattice phonons. This energy loss cannot be recovered and, therefore, decreases drastically the efficiency of solar cells due to the unavoidable "thermalization losses". In the following, we will briefly discuss some salient features of the thermalization process.

1. **Thermalization in molecules**
   Thermalization of excess kinetic energy in molecules can be described in the so-called *energy-configuration diagram*, $E(Q)$, as in Fig. 8.9. Here, the molecular energy E is plotted as a function of a multidimensional configuration coordinate Q describing the equilibrium atomic positions. For a bi-atomic molecule, Q would just be the distance between the two atoms. Close to the initial state $|i\rangle$, the energy has a minimum, so that $E(Q)$ can be approximated by a parabola. Molecular vibrations around the stable initial position can be accounted for by an equidistant ladder of a harmonic oscillator added to the electronic ground state. In a similar way, the metastable excited state $|f\rangle$ after an optical transition can be visualized by a second parabola, albeit shifted up in energy and with a different equilibrium configuration coordinate and a different spacing of harmonic vibrational sub-levels. According to the Franck–Condon principle, optically induced electronic transitions between the states $|i\rangle$ and $|f\rangle$ occur on a timescale much faster than possible changes of the molecular configuration Q, so they can be described by vertical arrows up or down in the $E(Q)$ diagram:
   Therefore, starting from the ground $|i\rangle$, an optical excitation will result in the occupation of an electronically and vibrationally excited level of state $|f\rangle$. If the excited molecule is in contact with a phonon bath (e.g. other molecules or a substrate), a first thermalization chain will occur by phonon transitions down the harmonic oscillator ladder of state $|f\rangle$ into the equilibrium excited state. From there, for example a second optical recombination event may occur by emission of a luminescence photon,[3] ending up in a vibrationally excited level of the ground state $|i\rangle$. From there, a second thermalization chain of phonon emissions will finally bring the molecule back to its ground state.
   Both thermalization chains give rise to the so-called *Stokes-Shift* between the excitation energy and the observed luminescence spectra, usually containing several vibrational side bands or *phonon replica*:

$$\hbar\omega_{ex} - \hbar\omega_{lum} = N\hbar\Omega \tag{8.2.21}$$

[3] Also called fluorescence or phosphorescence, depending on the lifetime of the excited state.

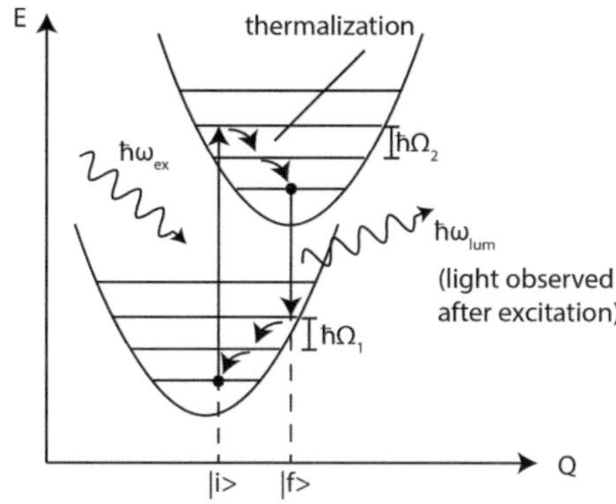

Fig. 8.9 Energy-configuration diagram of optical transitions and thermalization in a molecule. ($|i\rangle$: initial state, ex: excitation, $|f\rangle$: final state, lum: luminescence). See text for further details

2. **Thermalization in semiconductors**
In the case of semiconductors as absorber materials, thermalization is best discussed in the context of the usual $E(k)$ band diagram, where k is the wavevector of electrons and holes in the conduction band and valence bands, respectively (Fig. 8.10). For most inorganic semiconductors used in photovoltaics, the electronic band structure is as qualitatively shown in Fig. 8.10. The conduction band (CB) consists of linear combinations of atomic s-orbitals, forming strong ssσ-bonds.[4] In the valence band (VB), two possible combinations of atomic p-orbitals are present: weak ppπ-bonds between p-orbitals perpendicular to the bond axis and strong ppσ-bonds between p-orbitals parallel to the bond axis, as already discussed for C=C double bonds in Chap. 7. Electronic bands resulting from stronger bonds have a larger bonding-antibonding splitting and, therefore, a stronger curvature of $E(k)$ at the band extrema. This leads to a corresponding difference in the effective masses m^* of electrons and holes, defined by:

$$m^* = \hbar^2 \left(\frac{\partial E^2}{\partial k^2} \right)^{-1} \tag{8.2.22}$$

As a consequence, holes in the top-most valence of direct semiconductors have a much higher effective mass than electrons in the lowest conduction band. As can be seen in Fig. 8.10, this gives rise to the fact that, following a direct optical excitation, electrons in the conduction band have much more kinetic energy than holes in the valence band. As a consequence, thermalization occurs mainly by the relaxation of "hot" electrons back to $k = 0$ via a phonon emission cascade along the $E(k)$ band. In comparison, much less kinetic energy is lost by the thermalization of "warm" holes in the heavy hole valence band.
The most efficient phonon emission from hot electrons in compound semiconductors such as GaAs occurs via longitudinal optical (LO) phonons close to the

[4] xxσ- and xxπ- bonds denote whether the participating x-orbitals of the two bonding partners have rotational symmetry around or perpendicular to the bond axis, respectively.

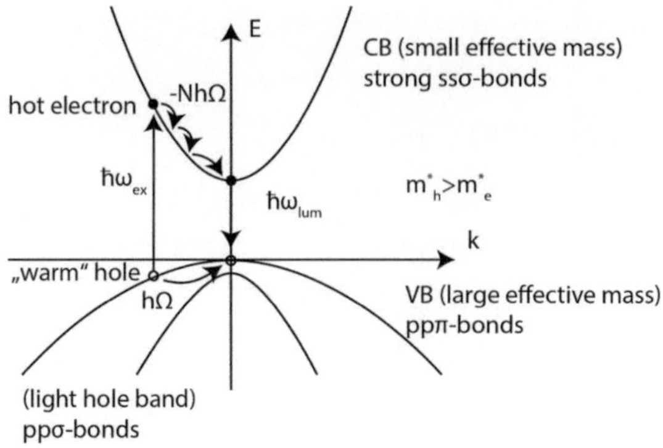

Fig. 8.10 Thermalization of optically excited electrons and holes in the lowest conduction band (CB) and the highest valence bands (VB), respectively. See text for details

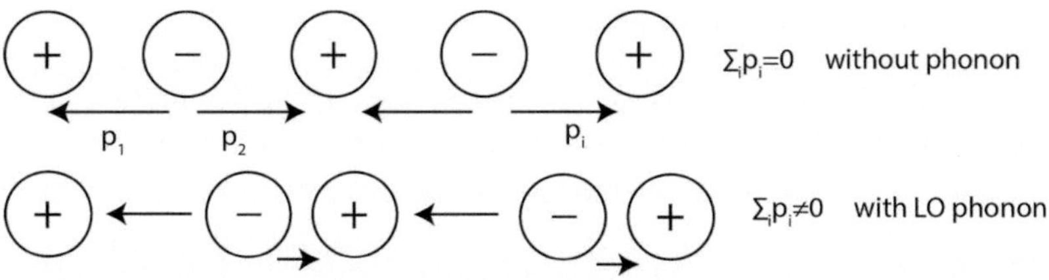

Fig. 8.11 Atomic positions and resulting dipole moments p_i per unit cell in a partially ionic compound semiconductor: at rest (top row) and with the displacement caused by a long wavelength ($k \approx 0$) longitudinal optical phonon (bottom row)

center of the Brillouin zone at $k \approx 0$, i.e. very long phonon wavelengths (the "Fröhlich" interaction). As shown in Fig. 8.11, the alternating positively and negatively charged atoms along a lattice direction give rise to nearest-neighbor dipole moments $\mathbf{p}_i$ with alternating directions. For an unperturbed atomic lattice with all atoms at rest, the macroscopic sum over these dipole moments is zero. In the presence of a long wavelength LO-phonon, however, the relative displacement of positively and negatively charged atoms will add up to a finite macroscopic polarisation, $\mathbf{P} = (1/v)\Sigma_i \mathbf{p}_i$. This macroscopic polarization $\mathbf{P}$ is equivalent to a macroscopic electric field $\mathcal{E}$, which changes direction according to the vibrational frequency of the LO-phonon and, thus, scatters charge carriers very effectively.

8.2.5 Recombination

After generation with rate G and before extraction, electrons and holes can recombine with the **recombination rate** R according to the overall rate equation:

$$\boxed{\frac{\mathrm{d}n}{\mathrm{d}t} = \frac{\mathrm{d}p}{\mathrm{d}t} = G - R} \tag{8.2.23}$$

Here, $n(t) = n_0 + \Delta n(t)$ is the is the total electron density, consisting of the equilibrium density n_0 due to doping or thermal excitation, and the time-varying optically excited carrier density $\Delta n(t)$. Similarly, the total hole density is given by $p(t) = p_0 + \Delta p(t)$.

Note that for the absorber layers of many solar cells one has a stronger p-doping with $p_0 > n_0$, so that holes are the *majority carriers*, whereas electrons are the *minority carriers*. This is chosen, because electrons in most cases have a smaller effective mass and, thus, a higher mobility μ and a larger diffusion constant D than holes:

$$\mu = \frac{v_{\mathrm{drift}}}{\mathcal{E}} = \frac{e\tau}{m^*}; \qquad D - \frac{kT}{e}\mu \tag{8.2.24}$$

Here, τ is the scattering time limiting the mobility. With a higher mobility, electrons can, thus, drift or diffuse a longer distance, so that the probability for extraction before recombination is increased.

Recombination can occur radiatively or non-radiatively:

Radiative recombination is the inverse process to optical excitation, however occurring after thermalization. Typical photoluminescence spectra of GaN layers are shown in Fig. 8.12. The radiative recombination probability is governed by the optical dipole matrix element $\langle i|ex|f\rangle$. Classically, the power emitted by an oscillating dipole is:

$$P = \frac{1}{6\pi\varepsilon_0}\frac{\omega^4}{c^3}e^2x^2 \tag{8.2.25}$$

where x is the electron-hole distance. The corresponding radiative recombination lifetime τ_{rad} is

$$\boxed{\tau_{\mathrm{rad}} = \frac{\hbar\omega}{P} = 6\pi\varepsilon_0\hbar\frac{c^3}{\omega^3e^2x^2}} \tag{8.2.26}$$

- τ_{rad} is particularly short for semiconductors with large band gaps, since $\hbar\omega \approx E_{\mathrm{gap}}$.
- Due to momentum conservation, τ_{rad} in indirect semiconductors is much larger than in direct semiconductors.
- A typical value for $\hbar\omega = 1\,\mathrm{eV}$ and an electron-hole distance $x \approx 1\,\mathrm{nm}$ (exciton radius) is $\tau_{rad} \approx 5 \times 10^{-9}\,\mathrm{s}$ (compared to $\approx 1\,\mu\mathrm{s}$ for indirect semiconductors).

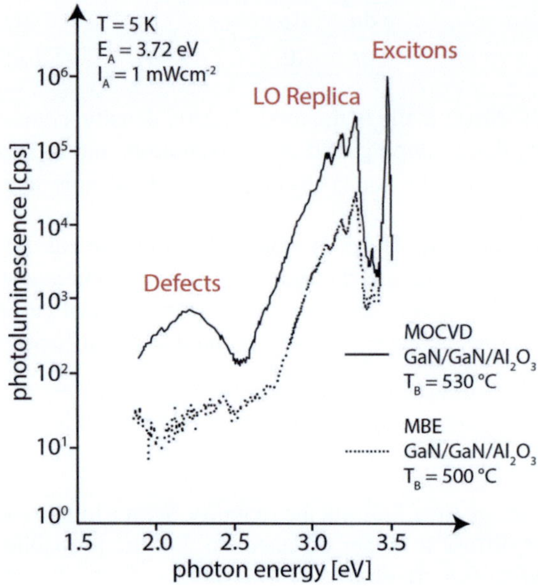

Fig. 8.12 Radiative recombination (photo-luminescence) at $T = 5\,\text{K}$ in GaN grown on Al_2O_3-substrates by two different methods: Metal-Organic Chemical Vapor Deposition (MOCVD) and Molecular Beam Epitaxy (MBE). Optical excitation occurred with a laser at an above-band gap energy of 3.72 eV with an intensity of $1\,\text{mW cm}^{-2}$. The spectra consist of a sharp excitonic peak close to the GaN band gap, recombination between shallow donor-acceptor pairs followed by a series of phonon replica, and radiative recombination at deep defects

Non-radiative recombination occurs via the emission of phonons. This can happen via:

- phonon cascades analogous to thermalization, if a high density of states is present in the band gap. However, this is normally not relevant for good solar cells.
- multi-phonon emission, i.e. the simultaneous emission of:

$$N = \frac{E_{gap}}{\hbar\Omega}$$

phonons in a process of order N. The corresponding probability W scales like:

$$W \propto \Delta_{e-phonon}^{N}$$

where $\Delta_{e-phonon} << 1$ is a measure for the electron-phonon coupling.

In an ideal semiconductor with no defect levels in the band gap E_{gap}, therefore $N >> 1$ holds. As a consequence, non-radiative recombination decreases with increasing band gap. Defect levels in the middle of the band gap generally act as very efficient *recombination centers*, as shown in Fig. 8.13.

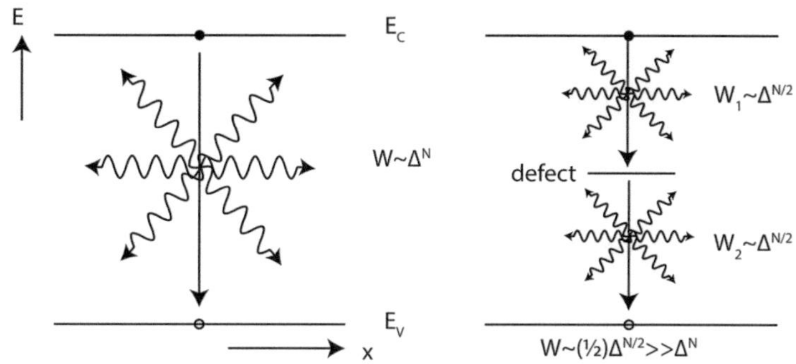

Fig. 8.13 Illustration of non-radiative recombination without and with midgap defects. In the right picture the recombination happens in two subsequent steps with lower order. Therefore, the overall probability W for recombination here is much larger compared to the left picture: $W \propto 1/2\Delta^{N/2} >> \Delta^N$

In the following, we will discuss the rate equations for some special recombination processes in more detail.

Monomolecular recombination: Here we have

$$\boxed{R = c_n \Delta n N_D} \qquad \boxed{R = c_p \Delta p N_D} \tag{8.2.27}$$

N_D is the volume density of defects acting as recombination centers, c_n the capture coefficient of the defects for electrons with the unit $[c_n] = \text{cm}^{-3}\text{s}^{-1}$.[5] Alternatively, one can also write for the capture coefficient $c_n = v_n \cdot \sigma_D$ with v_n, the electron velocity and σ_D, the capture cross section of defects. The resulting rate equation for monomolecular electron recombination (i.e. single electrons combine with a constant density of defects) is then:

$$\frac{\mathrm{d}n}{\mathrm{d}t} = G - R = G - c_n \Delta n N_D$$

In steady state ($\mathrm{d}n/\mathrm{d}t = 0$) this results in an excess density of electrons proportional to the generation rate and inversely proportional to the defect density:

$$\boxed{\Delta n = \frac{G}{c_n N_D}} \tag{8.2.28}$$

By introducing a carrier recombination lifetime τ_{rec} via $\tau_{rec} = \Delta n / R$, we obtain in steady state:

$$\boxed{R = \frac{\Delta n}{\tau_{rec}} \quad \Rightarrow \quad G - \frac{\Delta n}{\tau_{rec}} = 0} \quad \Rightarrow \quad \boxed{\tau_{rec} = \frac{1}{c_n \cdot N_D}} \tag{8.2.29}$$

[5] An analogous discussion holds for the capture of holes (index p).

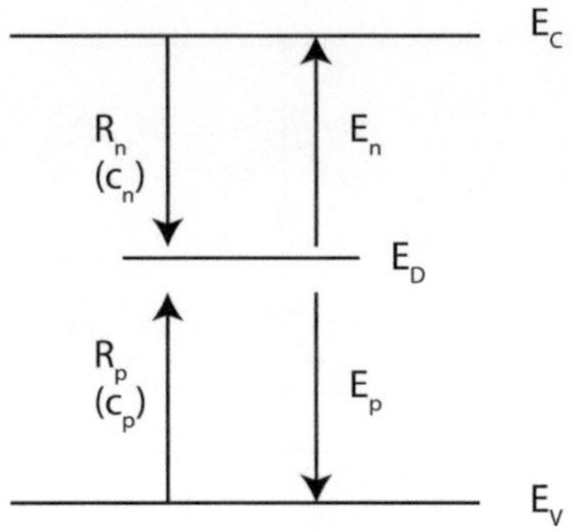

$$E_n \propto \exp\left(-\frac{E_C - E_D}{kT}\right)$$

$$E_p \propto \exp\left(-\frac{E_D - E_V}{kT}\right)$$

Fig. 8.14 Possible recombination rates R and thermal emission rates E between the valence and conduction band and a deep defect located at an energy E_D in the band gap. The emission rates are proportional to a Boltzmann term as shown on the right

A more detailed analysis also taking into accounts the thermal re-excitation of electrons trapped by defects into the conduction band as well as of trapped holes into the valence band was developed by Shockley, Read and Hall. Their model explicitly uses the principle of detailed balance between all recombination and excitation rates as shown in Fig. 8.14. In steady state ($R = E$), this gives:

$$R_{SRH} = \frac{N_D(np - n_0 p_0)}{c_p^{-1}\left[n + n_i \exp\left(\frac{E_D - E_i}{kT}\right)\right] + c_n^{-1}\left[p + n_i \exp\left(-\frac{E_D - E_i}{kT}\right)\right]} \tag{8.2.30}$$

Here, $n_i = p_i = \sqrt{N_V N_C} \exp\left(-\frac{E_{gap}}{2kT}\right)$ are the thermally excited intrinsic carrier densities without doping. N_V, N_C stand for the effective density of states of the valence and conduction band, for example:

$$N_C = \left(2/h^3\right)\left(2\pi m_e^* kT\right)^{3/2} \tag{8.2.31}$$

E_i is the position of the Fermi-level E_F in the intrinsic, undoped semiconductor, defined by:

$$n_i = N_C \exp\left(-\frac{E_C - E_i}{kT}\right) \tag{8.2.32}$$

Typical recombination centers in the most common semiconductors are:

- intrinsic defects such as vacancies, dangling bonds or dislocations
- charged impurities with a big Coulomb-like capture cross-section (esp. Fe, Ni, Au, …in Si).

Bimolecular recombination: Here, an optically excited electron in the conduction band recombines directly with a hole in the valence band. In the limit of strong excitation (i.e. $\Delta n >> n_0, \ \Delta p >> p_0$) this gives:

$$R = c^* \cdot \Delta n \Delta p$$

with $\Delta n = \Delta p$, so that in steady state:

$$0 = G - R = G - c^* \Delta n^2$$

This leads to a different dependence Δn on G as for monomolecular recombination:

$$\boxed{\Delta n = \sqrt{\frac{G}{c^*}}} \tag{8.2.33}$$

Here c^* is an effective capture coefficient for electrons by holes, and the electron recombination lifetime is:

$$\boxed{\tau_{rec} = 1/(c^* \Delta p)} \tag{8.2.34}$$

Characteristic for bimolecular recombination is the sublinear (square root) dependence of the excited carrier density on the generation rate.

Auger recombination: Here, the energy and momentum (wavevector) of a recombining electron-hole-pair is transferred through direct Coulomb interaction to a third electron or hole, which then rapidly returns by thermalisation to the ground state (Fig. 8.15).

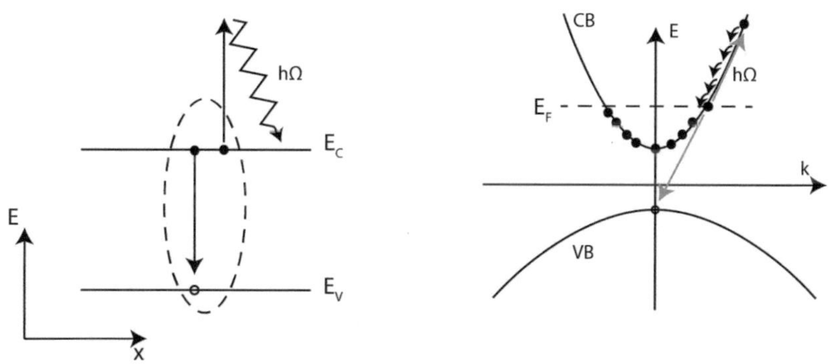

Fig. 8.15 Illustration of Auger recombination. The right picture shows a highly doped semiconductor, where the Fermi-level lies already within in the CB: this guarantees that there are enough electrons for this recombination process and momentum can be conserved

Auger recombination is a three-particle process via Coulomb interaction requiring either n_0 or p_0 to be sufficiently large. Depending on whether equilibrium holes or electrons are dominant, the Auger recombination rate is given by:

$$R = c_A \Delta n p_0^2 \quad \text{for strong p-doping, } p_0 >> \Delta p$$
$$R = c_A \Delta p n_0^2 \quad \text{for strong n-doping, } n_0 >> \Delta n$$

For steady state in heavily p-type semiconductors this leads to:

$$0 = G - R = G - c_A \Delta n p_o^2$$

This gives for the steady-state excess electron density:

$$\boxed{\Delta n = \frac{G}{c_A p_0^2}} \tag{8.2.35}$$

and

$$\boxed{\tau_{rec} = \frac{1}{c_A p_0^2}} \tag{8.2.36}$$

with a characteristic Auger coefficient c_A. In solar cells, Auger recombination may occur especially in highly doped contact regions with $n_0, p_0 \geqslant 10^{17}\text{cm}^{-3}$.

Surface recombination: For very high-quality absorber layers with little recombination in the bulk, recombination at surfaces or interfaces may become dominant. Most semiconductors have intrinsic surface or interface defects with energy levels in the band gap, which then can act as efficient recombination centers for photo-excited charge carriers coming sufficiently close to the surface or interface. For example, at the Si surface each Si atom is missing a fourth bonding partner, resulting in one so-called *dangling bond* orbital per atom (cf. Fig. 8.16 left). Neutral dangling bonds are occupied by a single electron carrying a spin (arrows in the left picture), which allows their detection via electron spin resonance. The dangling bonds form a defect band with an energetic position close to the middle of the bulk silicon band gap (center). Therefore, they can act as very efficient recombination centers for photo-excited carriers coming sufficiently close to the surface (Fig. 8.16 right). The neutral Si dangling bonds are *amphoteric* recombination centers: they can either capture a photo-excited electron from the conduction band (becoming negatively charged) or a hole from the valence band (becoming positively charged). The surface recombination is then completed by the effective Coulomb-attraction of an opposite charge carrier.

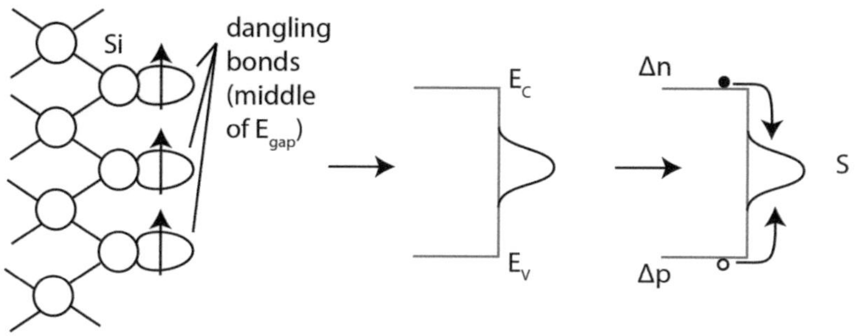

Fig. 8.16 Schematic illustration of surface recombination at a free unreconstructed surface of silicon. See text for details

The influence of the surface on the recombination lifetime τ_{rec} is described in terms of a *surface recombination velocity* S:

$$\boxed{S := \frac{R_{surface}}{\Delta n_{bulk}}} \tag{8.2.37}$$

The unit of the surface recombination rate is $[R_{surface}] = \text{cm}^{-2}\text{s}^{-1}$. The unit of the bulk photo-excited carrier density is $[\Delta n_{bulk}] = \text{cm}^{-3}$. Thus, the unit of S is indeed that of a velocity, cm s^{-1}. Typical values for S can vary largely between $S = 1\text{cm s}^{-1}$ for passivated surfaces (e.g. Si, where all surface dangling bonds are passivated by forming Si-H bonds with atomic hydrogen) to $S = 10^6\text{cm s}^{-1}$ (for bad, unpassivated surfaces). The upper limit of S is the thermal velocity v_{th} of minority carriers moving in the single direction towards the surface ($S < v_{th}$):

$$\frac{1}{2}m^* v_{th}^2 = \frac{1}{2}kT \qquad \Rightarrow \qquad S_{max} \approx 2 \cdot 10^7\text{cm s}^{-1}$$

As an example, Fig. 8.17 shows the relative contributions of different recombination processes to the overall carrier lifetimes in p-doped high quality crystalline silicon.

8.2.6 Separation and Extraction

Directly after optical excitation an electron and the corresponding hole first remain close together in a Coulomb-bound excitonic state. The excitonic binding energy is obtained analogous to that of the hydrogen atom (1 Rydberg):

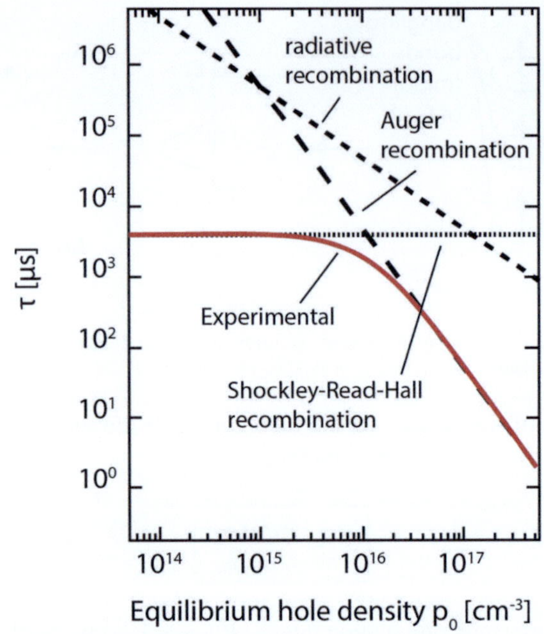

Fig. 8.17 Room temperature recombination lifetimes of electron minority carriers in p-doped solar-grade silicon with different equilibrium hole concentrations p_0. The experimentally observed lifetime (solid line) varies between 1ms and 1μs [2]. At low p-doping levels, it is dominated by monomolecular Shockley–Read–Hall recombination (dotted line), whereas Auger recombination dominates at high doping levels above 10^{16} cm^{-3} (dashed line). Due to the indirect nature and the low value of the silicon band gap, radiative recombination (fine dashed line) is always negligible

$$E_{bind} = \frac{e^4 m^*_{red}}{8h^2\varepsilon^2\varepsilon_0^2}\frac{1}{n^2}; \qquad n = 1, 2, \ldots; \qquad \text{with} \quad \frac{1}{m^*_{red}} = \frac{1}{m^*_e} + \frac{1}{m^*_h}$$

$$E_{bind}(n = 1) = 1\text{Ryd}\frac{m^*_{red}}{m_e}\frac{1}{\varepsilon^2} \tag{8.2.38}$$

With 1 Ryd = 13.959 eV this typically gives $E_{bind} \approx 10$ meV due to the reduced effective mass and high dielectric constant ($\varepsilon \approx 10$) in most semiconductors. Therefore, at room temperature phonons are already sufficient to dissociate the exciton and separate the electron and the hole through electron-phonon interaction (Fig. 8.18).

A more efficient separation occurs in internal electric fields e.g. in p/n-junctions, p/i/n-junctions, metal/semiconductor junctions or with an applied external bias voltage. In the $E(x)$-diagram, such an internal field leads to an inclination of both band edges (Fig. 8.19). The required field can be estimated from the potential energy drop across the radius R_1 of the excitonic ground state:

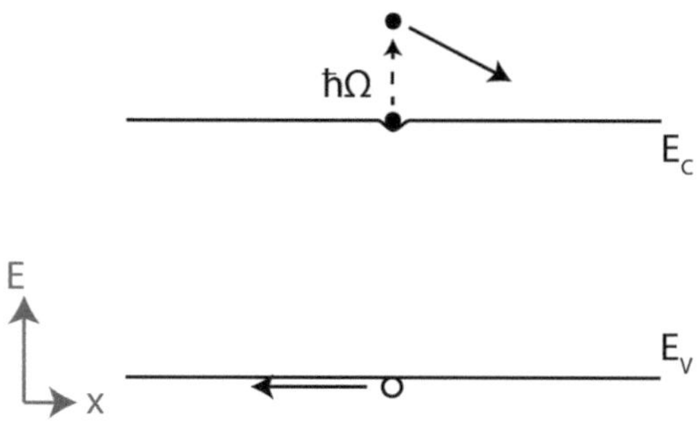

Fig. 8.18 Thermal separation of an optically excited electron-hole pair. The excitonic binding energy is indicated as a small dip in the conduction band when electron and hole are close together. The dashed arrow indicates excitation of the electron by absorption of a phonon followed by exciton dissociation

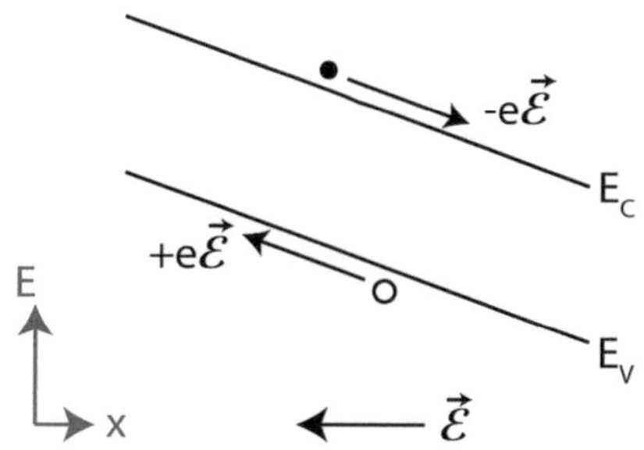

Fig. 8.19 Linear spatial inclination of the valence and conduction band edges due to a constant electric field, causing the drift of electrons and holes in opposite directions

$$e\mathcal{E}R_1 \geq E_{bind}(n=1)$$
$$\mathcal{E} = \frac{E_{bind}}{eR_1}$$
$$\text{with} \qquad R_1 = 4\pi\varepsilon\varepsilon_0\frac{\hbar^2}{e^2 m^*_{red}}$$

For a typical value of $R_1 \approx 5\,\mathrm{nm}$ and a binding energy of 10 meV, this gives an electric field of $\mathcal{E} \geq 20000\,\mathrm{V\,cm^{-1}}$.

In organic solar cells, excitonic binding energies are much larger ($\approx 0.5\,\mathrm{eV}$) because of the lower dielectric constant. Therefore, thermal exciton dissociation is no longer sufficient. An efficient separation of optically excited electrons and holes is then achieved at a suitable internal hetero-junction, as shown in Fig. 8.20 below.

The transport of separated electrons and holes to the contacts, where they finally are extracted, either occurs by diffusion (in a concentration gradient) or drift (in electric fields). This gives e.g. for the current density of electrons in one dimension:

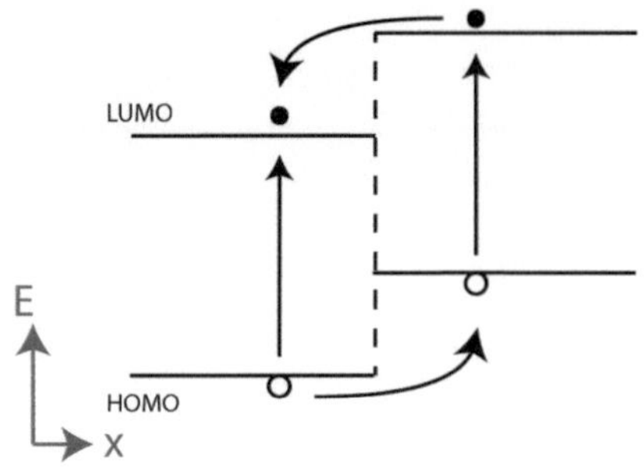

Fig. 8.20 Hetero-junction between two organic semiconductors with shifted HOMO and LUMO energies. After optical excitation, electrons will preferentially be trapped in the material with the lower LUMO energy, while holes will collect preferentially in the material with the higher HOMO energy

Table 8.2 μ_n, D_n and L_n for electrons at room temperature in different materials. For L_n, a recombination lifetime of $1\,\mu s$ has been assumed

Typical values	μ_n (300 K) [cm^2/(V s)]	D_n (300 K) [cm^2/m]	L_n ($\tau_{rec} \approx 10^{-6}$s) [μm]
crystalline Si	1000	25	50
amorphous Si	0.1	2.5×10^{-3}	0.5
polymers	10^{-3}	2.5×10^{-5}	0.05

$$j_n = env = \underbrace{en\mu_n\mathcal{E}}_{drift} + \underbrace{eD_n\frac{\partial n}{\partial x}}_{diffusion} \tag{8.2.39}$$

$$\text{with } \mu_n = \frac{e}{m_e^*}\tau \qquad \text{and } D_n = \frac{k_BT}{e}\mu_n$$

Thus, large mobilities μ_n, μ_p and the corresponding large diffusion coefficients D_n, D_p are required to allow for a good carrier extraction. During their recombination lifetime τ_{rec} charge carriers can diffuse over a distance L_n or L_p before recombining:

$$\boxed{L_{n,p} = \sqrt{D_{n,p}\tau_{rec}}} \tag{8.2.40}$$

Typical values of μ_n, D_n and L_n for electrons at room temperature in different material classes are listed in the Table 8.2. For $L >> 1/\alpha$ ($1/e$ absorption depth of light), solar cells can rely on diffusion for carrier extraction, otherwise drift in an internal field is required.

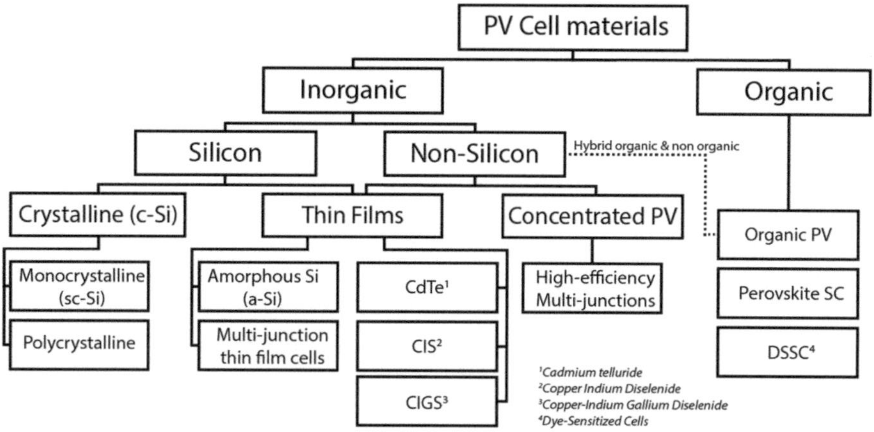

Fig. 8.21 Most common materials used for solar cell production. The work horse of photovoltaics with a market share above 95% is silicon in the form of monocrystalline or polycrystalline wafers. In thin film technology, compound semiconductors with a direct band gap such as CdTe or copper-indium-diselenide (CIS) are dominant. Multijunction cells with up to six stacked junctions predominantly use high quality III-V semiconductor alloys. Due to their high cost, multijunction cells are mainly used in space applications or in concentrated photovoltaics, where the cell sizes can be much smaller. The second class of solar cells is based on purely organic materials or combines organic and inorganic components in hybrid technologies

8.3 Types of Solar Cells

The history of modern solar cells for energy conversion basically started around 1955 with a brief note by Chapin, Fuller and Pearson of Bell Telephone Laboratories in the Journal of Applied Physics about a p/n-junction device made from crystalline silicon with an overall solar conversion efficiency of 6%. This was much more than what had been observed for at that time commercially available semiconductor-based photodetectors or thermoelectric devices (see Chap. 9) of below 1%. The authors also estimated that an optimized Si solar cell could have efficiencies around 22%, which is very close to the currently reported record efficiency of 26%. Since these early days, solar cells have experienced a very dynamic and successful development in terms of conversion efficiencies, industrial production and deployment as well as a cost reduction. At the same time, many other material systems have been investigated and developed for their use in photovoltaic converters (Fig. 8.21).

Today, many different materials and technological approaches are actively investigated for different purposes. However, only few of them currently contribute significantly to large scale solar energy conversion, notably crystalline silicon, CdTe and copper-indium-gallium-diselenide . One important physical parameter guiding the choice of materials is the band gap of the absorber layer, which has to be matched to the solar spectrum. If the band gap is too large, too few solar photons are absorbed and the electrical current density provided by the solar cell is small. If the band gap is too small, most energy is lost in unwanted thermalization losses and also the voltage provided by the cell is small. For single junction cells, there is a broad band

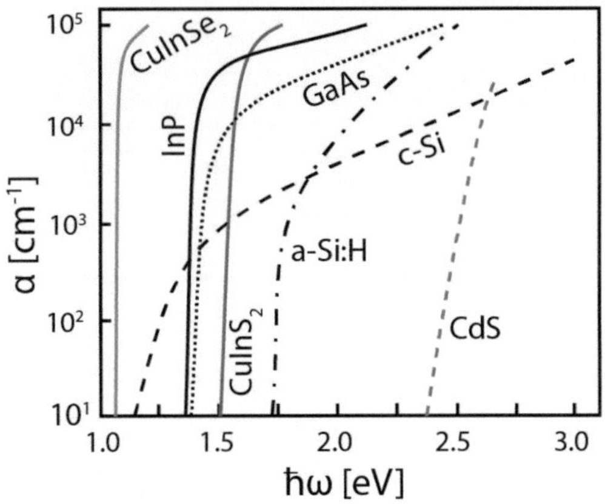

Fig. 8.22 Absorption coefficient spectra close to the band gap of some common absorber materials in inorganic solar cells

gap range between 1 and 1.6 eV where a maximum theoretical conversion efficiency around 30% could be reached. For multijunction cells, different absorber layers with carefully matched band gaps and thicknesses are used. Absorption coefficients of a small collection of materials with suitable band gaps are shown in Fig. 8.22.

A chart summarizing certified solar cell efficiencies for the different photovoltaic technologies is published and updated annually by the National Renewable Energy Laboratory. The 2022 version of this chart is reproduced below. Note that this chart only refers to best results obtained for small scale research cells. Mass-produced solar modules typically have efficiencies which are a third or more lower than these record efficiencies.

As can be seen, mature solar cell technologies such as monocrystalline silicon wafer cells or CIGS thin film cells have reached a plateau at around 25% efficiency. The highest efficiencies approaching 50% have been reached for multijunction cells using concentrated sunlight. More recent technologies with still significant improvements and fast learning curves have been organic solar cells since about 2000 and perovskite cells since about 2010. In the following we briefly discuss a few types of solar cells in more detail to introduce some important aspects of the underlying technology (Fig. 8.23).

8.3.1 Crystalline Si p/n Diffusion Cell

The basic band diagram of a crystalline silicon solar cell is shown in Fig. 8.24. It is based on mono- or poly-crystalline Si wafers with a typical size of $15 \times 15\,\mathrm{cm}^2$ and a thickness of 200–300 μm. This thickness is much larger than the value of $3/a$ required by optical absorption. Instead it is determined by the mechanical strength of the wafers in order to guarantee a sufficiently high yield in large scale automated production. Usually, the major part of the wafer forms the base, which is weakly p-type doped with an acceptor concentration N_A around 10^{15} cm^{-3}. Then optically excited electrons are the minority carriers and can diffuse freely through the entire base due to their higher mobility compared to holes as the majority carriers. In

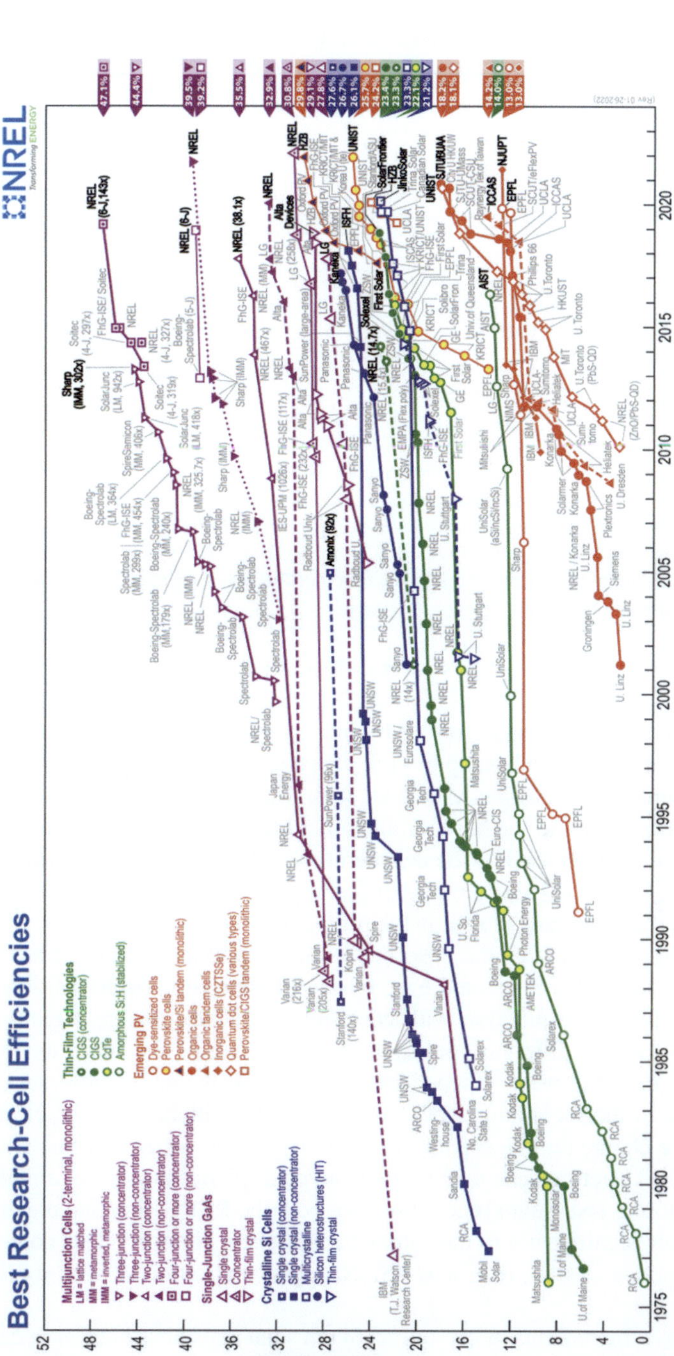

Fig. 8.23 Research solar cell efficiencies up to 2022. Permission for reproduction by NREL is gratefully acknowledged. The original graph can be found at: https://www.nrel.gov/pv/cell-efficiency.html

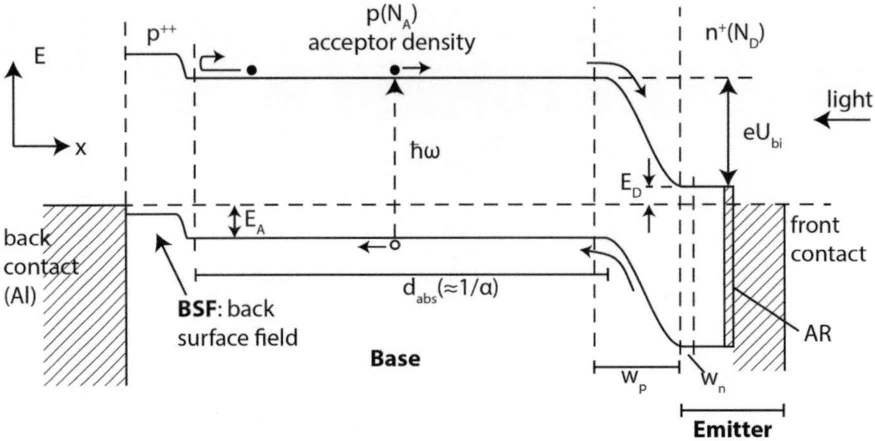

Fig. 8.24 Schematic band edge level diagram of a c-Si p/n diffusion solar cell. p^{++}, p and n^+ denote different doping levels with acceptors (density N_A) and donors (N_D). The corresponding dopant ionization energies E_A and E_D determine the position of the Fermi level in thermal equilibrium shown by the dashed horizontal line (without illumination). W_p and W_n are the widths of the space charge layers at the p/n-junction, across which the built-in potential eU_{bi} drops. AR stands for a suitable anti-reflection coating

high quality monocrystalline Si wafers diffusion lengths of more than 1 mm can be obtained.

The back contact is made from Al which is thermally diffused into the wafer. Since Al is a substitutional acceptor in Si, this creates a thin very highly doped (p^{++}) layer between the base and the Al contact, with two functions. On one hand, it insures a good ohmic contact to the base, on the other hand it causes a small upward jump in the band diagram caused by the degenerate p^{++}-doping. This creates the so-called *back surface field (BSF)* which repels diffusing electrons away from the back contact and attracts diffusing holes towards the p-type back contact.

On the opposite side of the absorber, a thin n-type layer called emitter with a high doping level (n^+) is created e. g. by in-diffusion of phosphorus donors. This results in a p/n-diode junction with corresponding space charge regions W_p and W_n. Because of the different doping levels, the extent of the space charge region is much larger in the weakly doped base, where also the generation rate of photo-carriers due to the incoming light (from the right) is highest. In the electric field of this space charge region, separation of photo-excited carriers occurs with a high efficiency, and electrons can drift to the front contact. Because of their high doping levels promoting fast Auger recombination, both the BSF region and the emitter do not contribute significantly to the solar cell efficiency. The top side is completed by an anti-reflection coating (e.g. a $\lambda/4$-Si_3N_4 layer) and a grid of screen-printed Ag contacts.

These basic ingredients of crystalline Si p/n solar cells have been varied and optimized over now almost 70 years. As an example, Fig. 8.25 shows a three-dimensional view of the so-called *PERL cell* (Passivated Emitter and Rear Locally diffused solar cell) developed at the University of New South Wales in Australia in 1998. This was

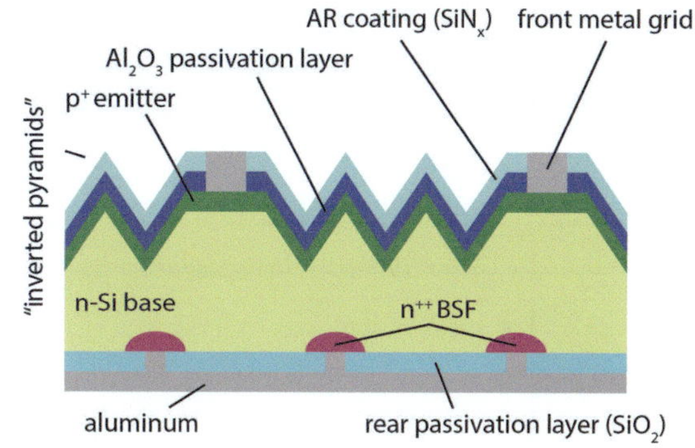

Fig. 8.25 Three-dimensional layout of a *PERL cell* (Passivated Emitter and Rear Locally diffused solar cell). See text for more details. Here, an n-type base instead of a p-type base has been used to avoid stability problems due to boron acceptors

the first Si p/n solar cell with a certified efficiency of 25% and made use of optical lithography to produce local p and n contacts within an otherwise passivated Si wafer surface covered by SiO_2. Lithography can also be used to create a regular inverted pyramid structure at the front surface by selective etching and was combined with an additional $\lambda/4$ layer as an efficient anti-reflection coating. The design of local backside contacts in combination with backside passivation is also referred to as *PERC cell* (Passivated Emitter and Rear Cell). Finally, the latest improvements in Si solar cells are based on the introduction of passivation layers using very thin hydrogenated amorphous silicon films on backside and front surfaces. These are called *HIT cells* (Heterojunction with Intrinsic Thin layer) and were introduced by Sanyo Electric Company around the year 2000. This type of crystalline Si cells currently holds the record in conversion efficiency of almost 27%.

To calculate the spatial profile of the conduction and valence band edges E_C and E_V one starts with the Poisson equation relating the local space charge density ρ with the divergence of the electrical field and the potential Φ:

$$\mathrm{div}\mathcal{E} = \frac{\rho}{\varepsilon\varepsilon_0}; \qquad \mathcal{E} = -\mathrm{grad}\phi \qquad \Rightarrow \qquad \mathrm{div(grad)}\phi = \Delta\phi = -\frac{\rho}{\varepsilon\varepsilon_o}$$

In doped semiconductors the local space charge density is given by the difference of ionized donor and acceptor densities: $\rho = N_D - N_A$.

- The built-in potential can be approximated by: $eU_{bi} \approx E_{gap} - E_D - E_A$
- The extent of the space charge regions W_p and W_n is obtained from:

$$W = \left(\frac{2\varepsilon\varepsilon_0}{e}\left(\frac{N_A + N_D}{N_A N_D}\right)\left(U_{bi} - 2\frac{kT}{e}\right)\right)^{1/2}$$

8.3.2 Metal-Insulator-Semiconductor (MIS)-Schottky Contact Cell

This is a second type of wafer-based diffusion cells using the electrical field generated by a top Schottky contact instead of a p/n-diode for carrier separation and extraction.

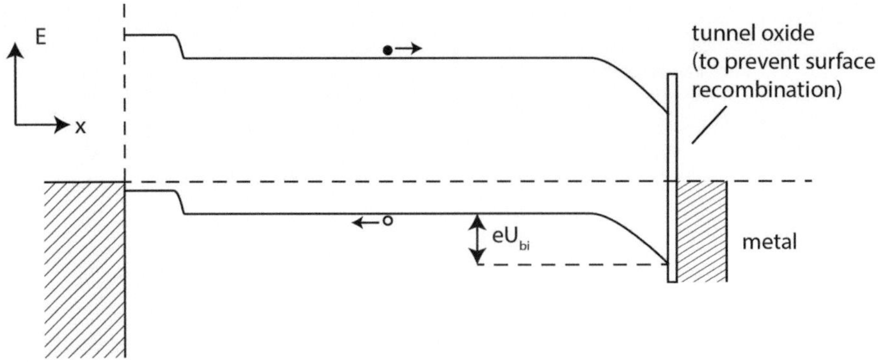

Fig. 8.26 Band level diagram of a silicon wafer-based Metal-Insulator-Semiconductor (MIS) solar cell. The n^+-doped emitter is replaced by a metallic Schottky contact, separated from the base by a thin tunnel oxide for surface passivation

Because of the missing n^+-emitter, it is easier to fabricate and requires a smaller thermal budget, however until now also has only achieved lower efficiencies. In such a MIS cell, a thin tunnel oxide or oxynitride is placed between the base and the top Schottky contact metal. The main purpose of the tunnel oxide is to prevent surface recombination. Otherwise, the band diagram and layout of a MIS cell resembles that of p/n-junction cells without emitter (Fig. 8.26).

8.3.3 Amorphous Si Thin Film Drift Solar Cells

Starting around 1980, thin films of hydrogenated amorphous silicon (a- Si:H) were investigated as an alternative to crystalline Si cells. This development was enabled by the then unexpected discovery that also this disordered form of silicon could be doped n- and p-type by the addition of phosphorus and boron, similar to its crystalline counterpart. A-Si:H can be deposited by Plasma-Enhanced Chemical Vapor Deposition (PECVD) from silane (SiH_4) at much lower temperatures around 250 °C on low-cost glass substrates coated with a contact layer using transparent conductive oxides such as ZnO, SnO_2, or indium-tin oxides (ITOs). In addition, the optical absorption coefficients of a-Si:H in the visible spectral range are much higher than those of crystalline Si (cf. Fig. 8.22), because the lack of crystalline periodicity in the amorphous network, removing the requirement of wavevector-conservation during an optical excitation and resulting in a "quasi-direct" optical band gap already mentioned above. This enabled an a-Si:H-based thin-film technology with absorber layer thicknesses below 1 μm, featuring much less material use and a much lower thermal budget. The major disadvantage of amorphous a-Si:H compared to crystalline Si on the other hand is the much lower electron and hole mobility (typically below $1\,cm^2\,V^{-1}s^{-1}$ at room temperature) due to strong carrier scattering by the disordered atomic network. This requires the presence of a sufficiently strong electrical field throughout the absorption layer to move photo-excited carriers by electrical drift to the contacts. Diffusion would be too slow to extract carriers before they recombine.

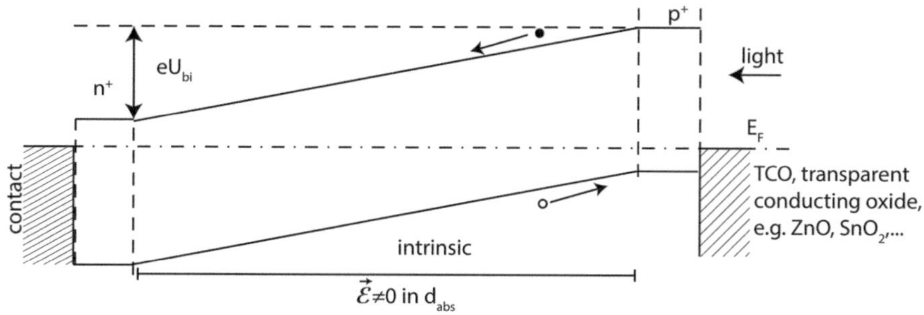

Fig. 8.27 Band level diagram of an amorphous silicon thin film drift cell

This motivates the layout of an amorphous Si solar cell as a p-i-n drift cell with a high electrical field in an undoped (intrinsic, "i") absorber layer. This field is created by strongly doped n^+ and p^+ extraction layers on opposite sides, followed by suitable metallic and transparent conducting oxide (TCO) contacts, as shown in Fig. 8.27.

For a typical absorber layer thickness of $d_{abs} = 0.5\,\mu m$ and a difference of the equilibrium Fermi-level positions in the n- and p-contact layers of 1 eV, the electrical field in the absorber can be estimated to about:

$$\mathcal{E} \approx \frac{U_{bi}}{d_{abs}} \approx \frac{1\,\mathrm{V}}{0.5\,\mu\mathrm{m}} = 2 \times 10^4\,\mathrm{Vcm}^{-1}$$

Illumination of the a-Si:H cell occurs through the glass substrate and the TCO layer at the p-contact side. This is required by the fact that at typical operating temperatures the mobility of holes in a-Si:H is much (about a factor of 100) lower than that of electrons, so that optical generation of excess holes should occur as close to the p-contact as possible. Although a-Si:H solar cells very quickly achieved conversion efficiencies above 10% at much lower costs than crystalline Si cells, their commercial acceptance was severely affected by the phenomenon of light-induced degradation (the so-called *Staebler–Wronski effect*). During long-term (several 1000 h operation of a-Si:H solar cells in normal sunlight, their conversion efficiency decreases by several percent (absolute) due to the optical generation of additional recombination centers. Similar degradation effects also plague other thin film technologies such as the organic or perovskite solar cells described below.

8.3.4 CdTe and Cu(In,Ga)Se_2 Compound Thin Film Solar Cells

Parallel to the development of thin film solar cells based on amorphous Si, also thin film cells using polycrystalline CdTe or $CuInGaSe_2$ (CIGS) chalcopyrite compound absorber layers were developed since 1980. Both absorber materials are direct semiconductors, allowing a reduced absorber thickness around 1 μm. Due to the higher carrier mobility in the polycrystalline absorber layers, they make use of both, carrier drift and diffusion, for charge extraction. Their thermal budget is between that of

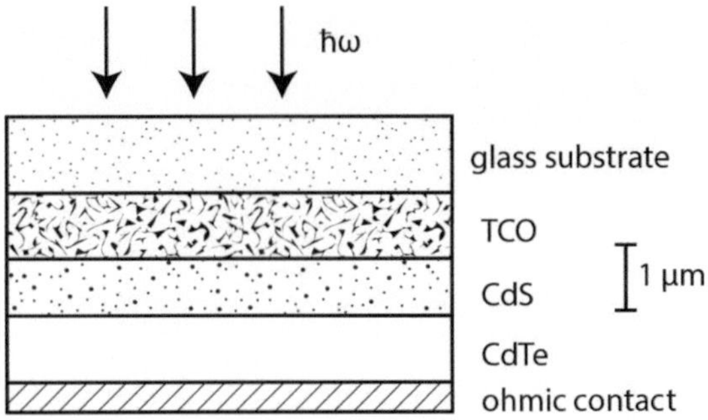

Fig. 8.28 Layer sequence of a CdTe compound thin film solar cell

a-Si:H and crystalline Si cells, depending on the specific deposition and processing methods used. Today, both compound thin film technologies have reached conversion efficiencies above 22% in the laboratory. Being a thin film technology, a glass sheet with a transparent conductive oxide is used as the substrate or superstrate of the cells (Fig. 8.28). In CdTe cells, a thin n-type CdS layer with a wider band gap of 2.4 eV is deposited as the n-contact and as a structural buffer layer. The p-type CdTe absorber ($E_{gap} = 1.45$ eV) is deposited next. The solar cell is then completed by an ohmic metal contact, e.g. Mo. In CIGS cells, basically only the CdTe absorber is replaced by Cu(In, Ga)Se_2. An alloy of In with Ga instead of pure In is used to increase the band gap of the absorber layer from around 1 eV to an optimum value around 1.4 eV. Because of the polycrystalline nature of the absorber, the conversion efficiencies of both types of cells are increased significantly by special grain boundary passivation treatments.

8.3.5 Dye-Sensitized Solar Cells (DSSC)

This thin film cell type was developed since 1990 first at EPFL and is also known as *Grätzel cell* after one of its inventors. It is inspired by early steps in photosynthesis for light absorption and carrier separation. Light absorption occurs in a molecular organic dye, which is dispersed on the large effective surface of nanostructured TiO_2. As shown in Fig. 8.29, a photo-excited electron is quickly transferred from the molecular LUMO to the conduction band of n-type TiO_2 and from there to the negative contact. The remaining hole in the HOMO of the dye molecule is then neutralized by an electron coming from a redox shuttle (e.g. iodine) in a liquid electrolyte and is transported to the TCO contact in form of the oxidized redox ion. There, it is reduced again by taking up an electron from the TCO. Today, dye-sensitized solar cells have achieved conversion efficiencies of up to 13%.

8.3.6 Organic Bulk Heterojunction Cell

Organic photovoltaics (OPV) is based on purely organic materials (molecular or polymeric) for light absorption and charge separation. As already discussed in the

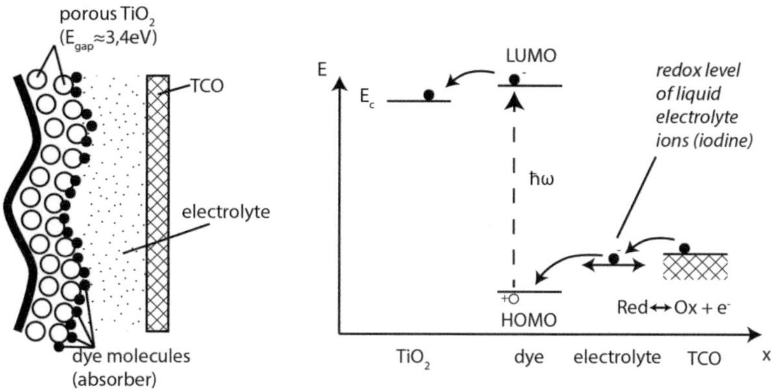

Fig. 8.29 Schematic structure and working principle of a dye-sensitized solar cell (DSSC)

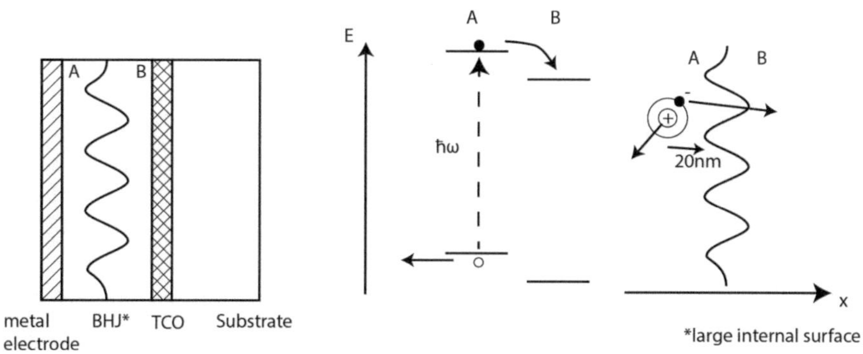

Fig. 8.30 Working principle of an organic bulk heterojunction cell. See text for details

beginning of this chapter, organic semiconductors are characterized by large exciton binding energies, so that abrupt heterojunctions between an electron donor material (A) and an electron acceptor material (B) are required for efficient carrier separation. After photo-excitation in A, the excited electron is transferred to B. Similarly, after photo-excitation in B, the hole is transferred to A. Thus, electrons are collected in B and holes in A. A second complication arises from the fact that excitons in organic materials also have a very low mobility for diffusion, which limits their diffusion lengths to a few 10 nm. To achieve a good compromise between a small average distance between A and B for efficient carrier separation and, at the same time, a sufficiently large combined thickness A + B for efficient photon absorption, the concept of a *bulk heterojunction (BHJ)* is used. Both components are intermixed on a nanoscale, forming a large effective interface as shown in Fig. 8.30, but still remain singly connected to allow a good transport of the separated carriers to the respective contacts. The development of organic solar cells started around the year 2000, leading to present record efficiencies of 18%.

A popular example of two organic components forming good bulk heterojunctions is shown in the Fig. 8.31.

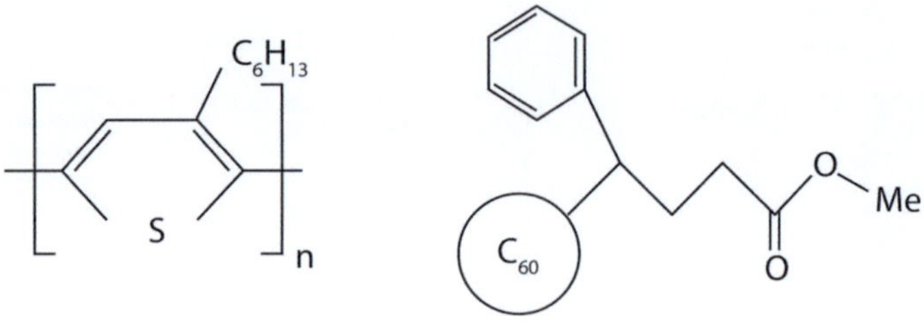

Fig. 8.31 Two components A and B for bulk heterojunctions: A = P3HT: Poly(3-hexylthiophene) (left). B = PCBM: [6, 6]-phenyl-C61-butyric acid methyl ester (right)

8.3.7 Perovskite Thin Film Solar Cells

The latest newcomer (2013) and the fastest rising star among the so-called emerging photovoltaic technologies with current record efficiencies above 25% is based on the material class of perovskites. Perovskites are ionic compounds with the stoichiometry ABX_3, where A and B are cations and X is an anion. The lattice structure of perovskites consists of AX_6 octahedra, eight of which surround a B cation as a cubic cage. Perovskites used for photovoltaics can be purely inorganic (such as $CsPbI_3$) or also contain organic cations. The best-known example for the latter is methylammonium lead iodide ($CH_3NH_3PbI_3$), which contributed a lot to the success of perovskite solar cells. Optical band gaps of perovskites are mostly direct and can be adapted easily to solar cell requirements by variation of the constituents. They also can be deposited with a high electronic quality by low-temperature deposition methods such as spin-coating from solution.

Figure 8.32 shows the basic layer sequence for solar cells using perovskite absorbers. As for almost all thin film solar cells, glass substrates covered by a transparent conductive oxide are used as the mechanical support and the optical window for the absorber layer. The perovskite absorber layer itself is sandwiched between other inorganic or organic layers to optimize carrier separation and collection by the final contact layers. Many of these interlayers have been copied from their earlier use in organic or dye-sensitized solar cells.

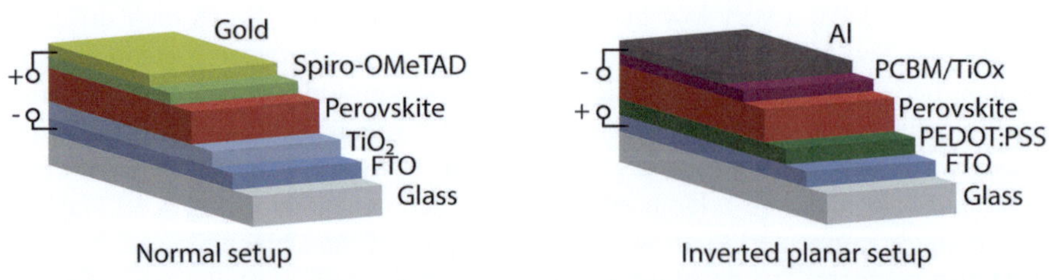

Fig. 8.32 Layer sequence of normal and inverted perovskite solar cells. FTO stands for the transparent conductive fluorine-doped tin oxide, TiO_2, PEDOT:PSS, PCBM and Spiro-OMeTAD are special electron or hole conductors helping to transport charge carriers to the final metallic contacts

8.4 I-U-Characteristics of Solar Cells

8.4.1 Ideal Diodes

The current-voltage-characteristics of an ideal pn-diode is obtained by solving the drift-diffusion equation for electrons and holes:

$$\mathbf{j}_n = -en\mu_n \mathcal{E} - eD_n \mathrm{grad}(n) \tag{8.4.1}$$

$$\mathbf{j}_p = ep\mu_p \mathcal{E} + eD_p \mathrm{grad}(p) \tag{8.4.2}$$

This gives (in the dark) (Fig. 8.33):

$$\boxed{j = j_s \left(\exp\left(\frac{eU}{\hat{n} k_B T} \right) - 1 \right)} \tag{8.4.3}$$

with the following quantities:

- the saturation current density for $U \to -\infty$:

$$j_s = e \left(\frac{D_n}{L_n} n_{0,p} + \frac{D_p}{L_p} p_{0,n} \right) \tag{8.4.4}$$

- the diffusion length $L = \sqrt{D\tau_{rec}}$
- $n_{0,p}$: equilibrium electron concentration on the p-side
- $p_{0,n}$: equilibrium hole concentration on the n-side

$$\boxed{n_{0,p},\ p_{0,n} \propto \exp\left(-\frac{E_{gap}}{k_B T} \right)} \tag{8.4.5}$$

- $\hat{n}$: the *ideality factor* with $1 \leq \hat{n} \leq 2$
  - $\hat{n} = 1$: no recombination in the space charge region W
  - $\hat{n} = 2$: complete recombination in the space charge region W (e.g. in a LED)

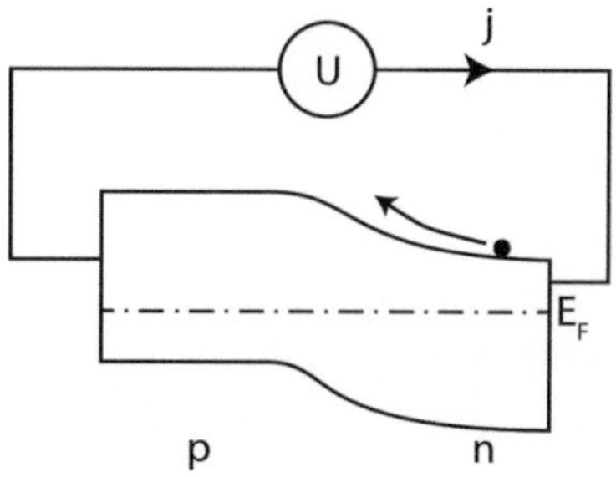

Fig. 8.33 Ideal circuit of a pn-junction in a semiconductor

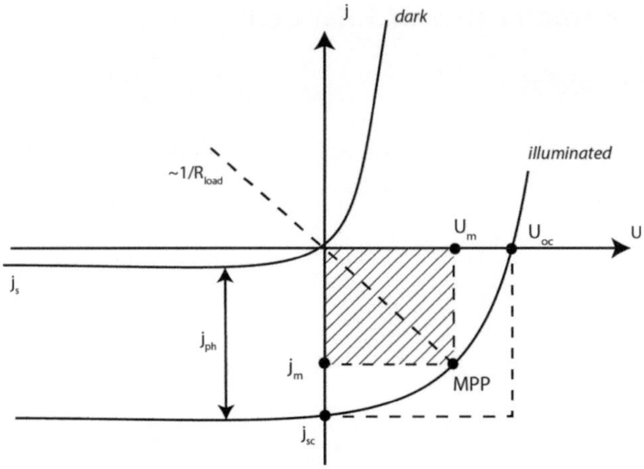

Fig. 8.34 Schematic current-voltage characteristics of an ideal pn-diode in the dark and under illumination. MPP is the maxiumum power point. See text for details

With illumination the photo-generated current density j_{ph} (e.g. $j_{ph} = e(\Delta n\mu_n + \Delta p\mu_p)\mathcal{E}$ for a drift cell) is subtracted from the dark I-U characteristics:

$$\boxed{j = j_s\left(\exp\left(\frac{eU}{\hat{n}k_BT}\right) - 1\right) - j_{ph}} \tag{8.4.6}$$

From this, the following important solar cell parameters can be deduced (Fig. 8.34):

- U_{oc}: the ***open circuit voltage***

$$j = 0 \quad \rightarrow \quad U_{oc} = \frac{nk_BT}{e}\ln\left(\frac{j_{ph}}{j_s} - 1\right) \tag{8.4.7}$$

- j_{sc}: the ***short circuit current density*** ($U = 0$)

$$U = 0 \quad \rightarrow j_{sc} = -j_{ph}$$

- the ***fill factor (FF)***:

$$FF = \frac{U_m j_m}{U_{oc} j_{sc}} \tag{8.4.8}$$

Here U_m and j_m are the voltage and current density in the *maximum power point (MPP)*, the point of operation where the maximum electrical power density can be drawn from the cell. This is determined via a suitable ohmic load resistor R_{load}

connected to the illuminated diode. With these parameters, the solar cell efficiency η can be written as:

$$\boxed{\eta = \frac{P_{out}}{P_{in}} = \frac{U_m j_m}{P_{in}} = \frac{FF \cdot U_{oc} \cdot j_{sc}}{P_{in}}} \tag{8.4.9}$$

where P_{in} is the incoming light power density ($[P_{in}] = \mathrm{W\,m^{-2}}$)

The open circuit voltage U_{oc} can also be described in terms of so-called *quasi-Fermi-levels* of electrons and holes. In the dark, the equilibrium carrier concentrations n_0 and p_0 are determined by the distance of the common Fermi-energy E_F from the band edges E_C and E_V:

$$n_0 = N_C \exp\left[-\frac{E_C - E_F}{kT}\right], \quad p_0 = N_V \exp\left[-\frac{E_F - E_V}{kT}\right]$$

Under illumination, both carrier concentrations increase to their non-equilibrium values: $n_0 \to n_0 + \Delta n = n$ and $p_0 \to p_0 + \Delta p = p$, with $\Delta n,\ \Delta p = G\tau_{rec}$. This allows the definition of the quasi-Fermi-levels $E_{F,n}$ and $E_{F,p}$ of electrons and holes according to:

$$n = n_0 + \Delta n =: N_C \exp\left[-\frac{E_C - E_{F,n}}{kT}\right]$$
$$p = p_0 + \Delta p =: N_V \exp\left[-\frac{E_{F,p} - E_V}{kT}\right]$$

The open circuit potential eU_{oc} is then given by the splitting of the two quasi-Fermi-levels as shown in the Fig. 8.35.

8.4.2 Real Diodes

The equivalent circuit of real pn-diodes (cf. Fig. 8.36) includes an ideal diode together with a parallel resistance R_P and a series resistance R_S: Both resistances modify the I-U characteristics of real diodes ($I = j \cdot A$) in the following way:

$$I = I_s\left(\exp\left(\frac{e(U - IR_s)}{\hat{n}k_B T}\right) - 1\right) + \frac{U - R_S I}{R_p} - I_{ph} \tag{8.4.10}$$

This equation takes into account that the voltage U across the ideal diode is reduced by the voltage drop $I \cdot R_S$ across the series resistor, and the same voltage drives an

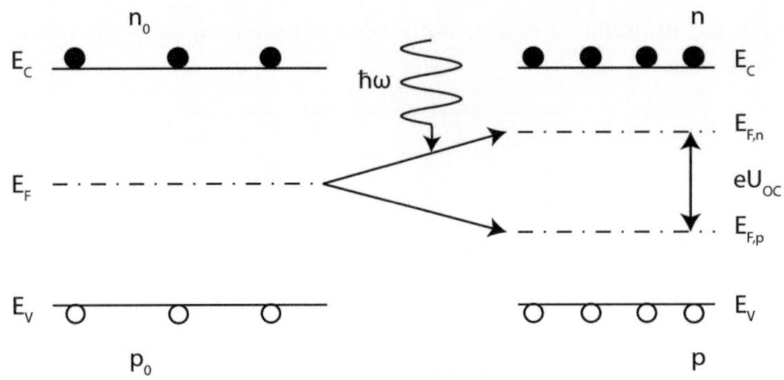

Fig. 8.35 Splitting of the quasi-Fermi-levels and resulting open circuit potential

Fig. 8.36 Equivalent circuit of real pn-diodes. Series resistances usually are caused by contacts and external leads, while parallel or shunt resistances are caused by conductive defects in the absorber layer (pin holes, impurities, etc.)

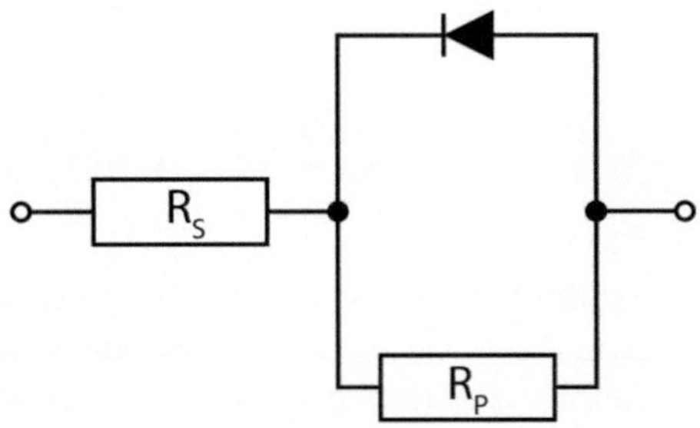

additional leakage current through R_P. R_S and R_P can be deduced from the slopes of the I-U curve at $U = 0$ and $U = U_{oc}$:

$$R_s = \left(\frac{\mathrm{d}I}{\mathrm{d}U}\right)^{-1}\Bigg|_{U_{oc}} \qquad R_s + R_p = \left(\frac{\mathrm{d}I}{\mathrm{d}U}\right)^{-1}\Bigg|_{U=0}$$

The series and parallel resistance of real diodes directly affect the solar cell efficiency in the following ways:

- Too large values of R_s decrease I_{sc} and FF
- Too small values of R_p decrease U_{oc} and FF
- in real devices with an area of $1\mathrm{cm}^{-2}$, R_s of a few Ω noticeably decreases the cell efficiency η, whereas only for $R_p \leq 100\Omega$ the efficiency is significantly affected.

8.5 Efficiency Limits of Single Junction Solar Cells

For the estimation of the maximum efficiency of an ideal pn-solar cell we make the following simplifying assumptions:

1. All photons with $\hbar\omega \geq E_{gap}$ are absorbed, all photons with $\hbar\omega \leq E_{gap}$ are transmitted. This means that there is no reflection.
2. No non-radiative recombination at defects or interfaces occurs.
3. No radiative recombination occurs or there is radiative equilibrium between the Sun and the solar cell. The latter is referred to as the *detailed balance* or **Shockley–Queisser limit** [5].
4. The solar cell is an ideal diode: $R_p = \infty;\ R_s = 0;\ \hat{n} = 1$

As discussed above, the solar cell efficiency per unit area is $\eta = (FF \cdot U_{oc} \cdot j_{sc})/P_{in}$ with P_{in} given by the solar spectrum:

$$P_{in} = \int_{\Omega}\int_{0}^{\infty} de_{\gamma}(\hbar\omega) \qquad \text{with} \tag{8.5.1}$$

$$\mathrm{d}e_{\gamma} = \frac{(\hbar\omega)^3}{4\pi^3\hbar^3 c_0^3}\left(\exp\left[\frac{\hbar\omega}{k_B T_{sun}}\right] - 1\right)^{-1} \mathrm{d}\hbar\omega d\Omega \tag{8.5.2}$$

Here, $T_{sun} = 5800\,\mathrm{K}$ and integration occurs over Ω according to the solid angle extended by the solar cell, or taking into account a possible concentration system.

The short circuit current density j_{sc} is only produced by photons with $\hbar\omega \geq E_{gap}$, and the maximum energy provided by these photons is eU_{oc}, due to thermalisation losses. Using the photon density of states and the thermal equilibrium occupation factor as derived at the beginning of this chapter, the maximum output power for a fill factor of $FF = 1$ would be:

$$P_{out} = \int_{\Omega}\int_{E_{gap}}^{\infty} eU_{oc} \cdot D_{\gamma,\Omega} f_{\gamma}(\hbar\omega)\mathrm{d}\hbar\omega\mathrm{d}\Omega = U_{oc} j_{sc}$$

This results in a maximum efficiency of a solar cell without hot carrier extraction (see below) of:

$$\eta_{max} = \frac{P_{out}}{P_{in}} = \frac{\int_{E_{gap}}^{\infty} eU_{oc}(\hbar\omega)^2\left(\exp\left[\frac{\hbar\omega}{k_B T_{sun}}\right] - 1\right)^{-1} d\hbar\omega}{\int_0^{\infty}(\hbar\omega)^3\left(\exp\left[\frac{\hbar\omega}{k_B T_{sun}}\right] - 1\right)^{-1} d\hbar\omega}$$

Thus, η_{max} depends strongly on the energy gap E_{gap} of the absorber because:

- $\eta_{max} \to 0$ for $E_{gap} \to 0$ ($U_{oc} < E_{gap} \to 0$)
- $\eta_{max} \to 0$ for $E_{gap} \to \infty$ (no adsorption, so that $j_{sc} \to 0$)

Under the simplifying assumptions made so far, there is a band gap optimum for the maximum efficiency with a value of $\eta_{max} = 42\%$ for $E_{gap} = 1.1\,\text{eV}$, assuming that $E_{oc} \approx E_{gap}$. Thus, Si with a room temperature band gap of $E_{gap} = 1.1\,\text{eV}$ would be the ideal absorber.

For a more accurate estimate of the efficiency limits, however, we also have to take into consideration the influence of a realistic $U_{oc} < E_{gap}$ and a realistic $FF < 1$. As discussed above, U_{oc} is determined by the splitting of the quasi-Fermi-levels $E_{F,n}$ and $E_{F,p}$ under illumination and depends on the generation rate G, but also on recombination processes and e.g. donor and acceptor levels of the pn-diode in the junction. This lowers eU_{oc} to typically around 70% of E_{gap}. A more accurate quantitative discussion of the dependence of the open-circuit-voltage on the band gap of the absorber is very complicated and would go beyond the scope of this book.

A second challenge is to estimate the fill factor of a solar cell in the maximum power point assuming an ideal diode current-voltage characteristics with known values of U_{oc} and I_{sc}. A starting point is the maximum power condition, leading to the following relations for the related voltage and current, U_m and I_m:

$$\mathrm{d}(IU)\Big|_m = 0 = U\mathrm{d}I + I\mathrm{d}U \quad \Rightarrow \quad \frac{\mathrm{d}I}{\mathrm{d}U}\Big|_m = -\frac{I}{U}\Big|_m$$

Inserting this into the ideal diode (8.4.3) gives:

$$\frac{\mathrm{d}I}{\mathrm{d}U}\Big|_m = I_s \frac{e}{kT}\exp\left(\frac{eU_m}{kT}\right) = -\frac{I}{U}\Big|_m \tag{8.5.3}$$

Moreover, for ideal diodes we have (Fig. 8.37)

$$I_m = I_s\left(\exp\left(\frac{eU_m}{kT}\right) - 1\right) - I_{sc}$$

$$\frac{I_{sc}}{I_s} = 1 - \exp\left(\frac{eU_{oc}}{kT}\right)$$

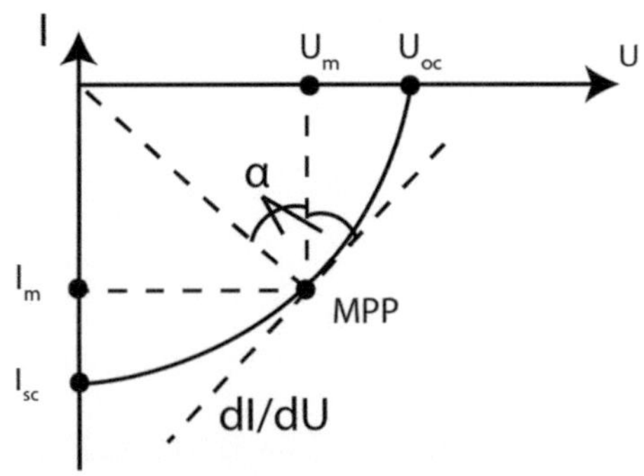

Fig. 8.37 Relation between the ratio I_m/U_m of current and voltage in the maximum power point of a solar cell and their derivate $\mathrm{d}I_m/\mathrm{d}U_m$

Inserting this into the above equation for the maximum power point results in:

$$U_m = -\frac{I_m}{I_s}\frac{kT}{e}\exp\left(-\frac{eU_m}{kT}\right) = \frac{kT}{e}\left(\exp\left(\frac{e(U_{oc}-U_m)}{kT}\right)-1\right)$$

This equation can be solved numerically to obtain U_m and I_m for a given U_{oc} and I_{sc} and, thus, the corresponding fill factor FF. In a good approximation one gets for ideal diodes:

$$FF \approx \frac{\frac{eU_{oc}}{kT} - \ln\left(1+\frac{eU_{oc}}{kT}\right)}{1+\frac{eU_{oc}}{kT}} \tag{8.5.4}$$

This leads to values of $FF \approx 80\%$ for optimized solar cells.

Some additional remarks concerning conversion efficiencies of real solar cells:

- Emission due to radiative recombination can usually be neglected if exciton separation is fast enough
- Non-radiative recombination causes the largest deviation from the maximum efficiency η_{max}
- Concentration of light leads to higher efficiencies due to an increase of j_{sc}, resulting in a corresponding increase of U_{oc} due to a larger splitting of the electron and hole quasi-Fermi-levels. For a concentration factor C this ideally gives:

$$U_{oc} \propto \frac{kT}{e}\ln\left(\frac{j_{sc}}{j_s}\right) \rightarrow U_{oc} = \frac{kT}{e}\ln\left(\frac{C\cdot j_{sc}}{j_s}\right)$$

  Thus, the efficiency of concentrated solar power (CSP) cells is larger.
- For strong concentration the temperature of the cell increases, which has to be prevented by active cooling. Otherwise, the dark saturation current j_s increases exponentially, decreasing U_{oc} correspondingly.

8.6 Increasing Solar Cell Efficiencies

The ideal conversion of energy from the Sun to energy on Earth would be subject to the thermodynamic Carnot efficiency limit defined by the absolute temperatures on the surfaces involved:

Table 8.3 Different approaches to increase the expected maximum conversion efficiencies η_{max} of solar cells

Expected η_{max} (%)	Approach
31	Single p/n junction without reflection and recombination
39	Including optical down-conversion
44	Tandem solar cell
46	Impact ionization (carrier multiplication)
49	Optical up-conversion
51	Triple solar cell
54	Thermal conversion, thermo-photovoltaics, thermionics
68	"hot carrier" cells, N$\rightarrow \infty$ multijunctions

$$\eta_{Carnot} = \frac{5800\,\text{K} - 300\,\text{K}}{5800\,\text{K}} = 95\%$$

Of course, this is not practical and the more realistic efficiency limit for single absorber solar cells is the Shockley–Queisser limit of about $\eta = 31\%$. Higher efficiencies would require circumventing the main loss mechanisms such as thermalisation and recombination. Indeed, several approaches exist or have been proposed to overcome the Shockley–Queisser limit by using multi-absorber solar cells or so-called "Third generation" solar cell concepts as listed in Table 8.3.

In the following, we will briefly sketch some examples for these approaches. A more detailed treatment of the corresponding research activities can be found in the related specialized literature.

8.6.1 Down Conversion

Optical down-conversion makes use of specific dye molecules, color centers or quantum dots in a transparent layer in front of a solar cell, to convert high energy photons with $\hbar\omega > 2E_{gap}$ into two photons with $\hbar\omega \approx E_{gap}$, where E_{gap} is the band gap energy of the subsequent solar cell absorber layer. This reduces the thermalization losses for high energy photons compared to their direct absorption in the absorber layer, thus increasing the overall efficiency (Fig. 8.38).

8.6.2 Tandem Cells

Tandem solar cells have been actively investigated since the advent of thin-film solar cells around 1980. They combine two single junction solar cells with different

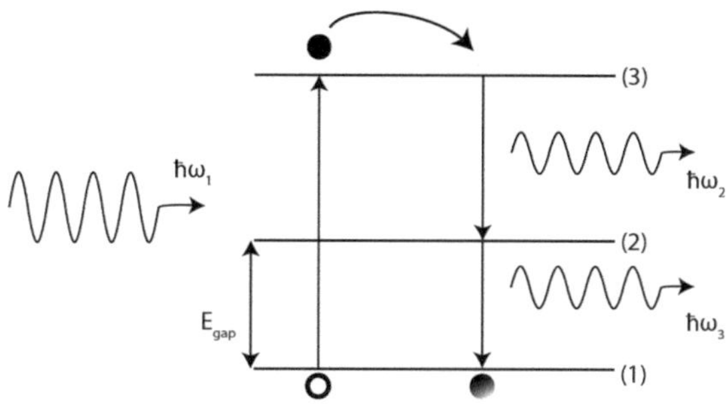

Fig. 8.38 Schematic electronic level sequence involved in optical down-conversion of high energy photons

optical band gaps E_{gap} placed optically and electrically in series. A top cell with larger band gap absorbes higher energy photons, while lower energy photons are subsequentially absorbed by the bottom cell with lower band gap. This theoretically allows a significant reduction of thermalization losses compared to a single junction cell, but requires a careful optimization of thickness and band gap of the two absorber layers for maximum efficiency under varying illumination conditions. Since tandem solar cells require the same illuminated area but reach a higher efficiency than single junction cells, they potentially provide a significantly lower system cost in real applications. As shown in Fig. 8.39, there are two possibilities two combine top and bottom cells electrically. In a three-terminal combination, a transparent conductive intermediate layer provides a third contact, which allows to operate both cells separately at their maximum power points. Most current approaches, however, are based on a two-terminal combination, so that the same current $I = I_1 = I_2$ flows through both cells, while the two cell voltages add up to a single external voltage $U = U1 + U2$. In order to realize this series interconnection, a dedicated tunnel junction has to be added as an additional element (see below).

For two-terminal tandem solar cells operated under $AM0$ conditions, theoretical efficiencies of up to 44% have been calculated for the combination of a top cell absorber with a band gap of 1.9 eV and a bottom cell absorber with a band gap of 1 eV (Fig. 8.40). These strict band gap limitations can be relaxed significantly when accepting reduced efficiency limits of about 40%.

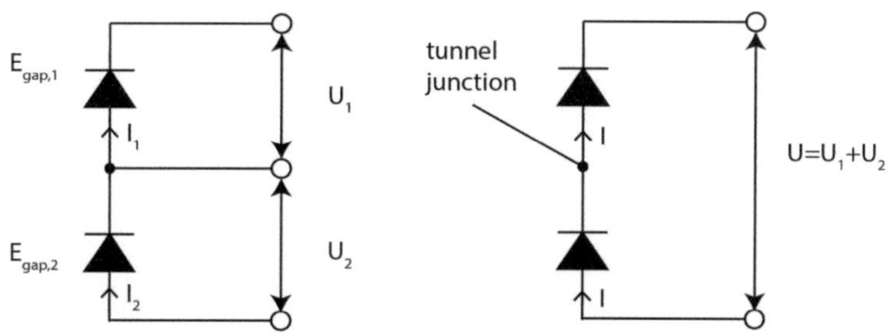

Fig. 8.39 Three-terminal and two-terminal electrical connections for tandem solar cells

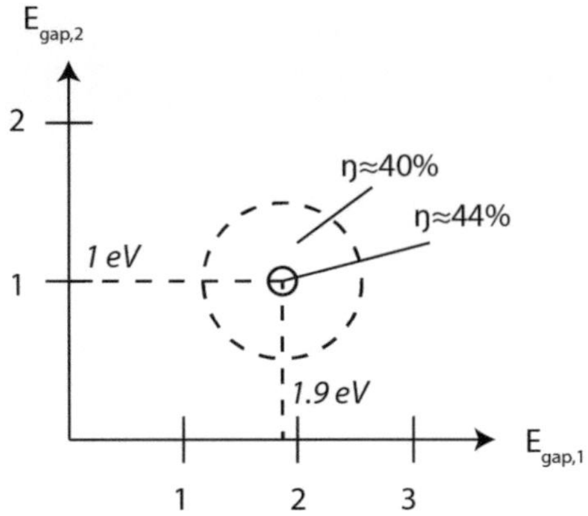

Fig. 8.40 Theoretical maximum efficiencies of two-terminal tandem solar cells under $AM0$ illumination for different combinations of top and bottom absorber band gaps

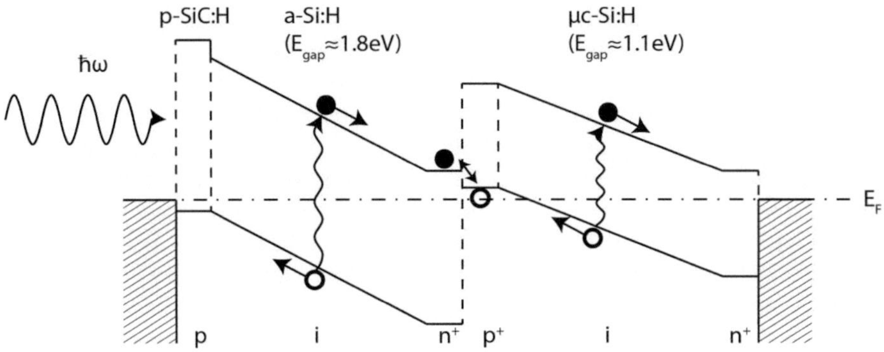

Fig. 8.41 Band level diagram of a thin film tandem solar cell combining an amorphous a-Si:H top cell with a microcrystalline μc-Si:H bottom cell. A p-type amorphous Si-C alloy layer acts as a transparent top contact and optical window. The electrical interconnect between both cells is achieved by a highly doped p/n tunnel junction

As a recent example, Fig. 8.41 shows the band level scheme of a so-called 2nd generation "micromorph" tandem thin film Si solar cell combining an amorphous Si top cell and a microcrystalline Si bottom cell. Both materials can be produced with a low thermal budget by plasma-enhanced chemical vapor deposition on low-cost glass substrates, are close to the ideal band gap combination of tandem solar cells and have achieved efficiencies of about 15%. Other current examples explore perovskite-based top cells combined with bulk wafer silicon solar cells or CIGS thin film cells.

For the electrical interconnection between two cells in a multi-junction solar cell, an efficient loss-less recombination between electrons (holes) of the top cell with holes (electrons) of the bottom cell needs to be realized. This is achieved by a dedicated tunnel junction involving heavily doped n^+ and p^+ layers. The doping

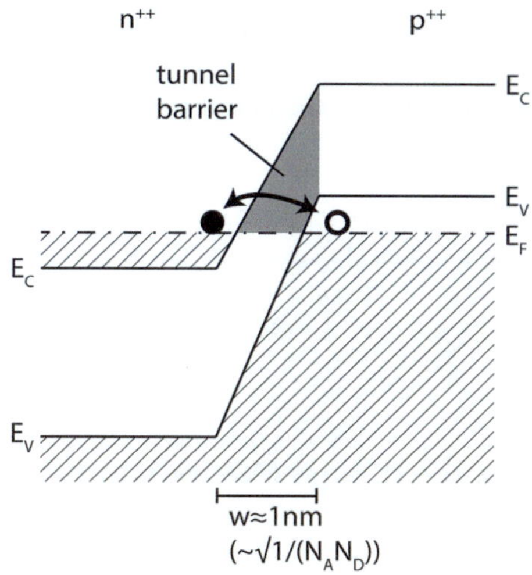

Fig. 8.42 Details of the band level alignment at a n^{++}/p^{++} tunnel junction between two sub-cells of a multi-junction solar cell. Due to the high doping levels, the Fermi energy E_F lies within the conduction band on the n-side and the valence band on the p-side (degenerate doping), allowing access to a large density of empty states on both sides. In addition, the extents of the space charge regions W are small enough to allow efficient carrier tunnelling through the junction barrier

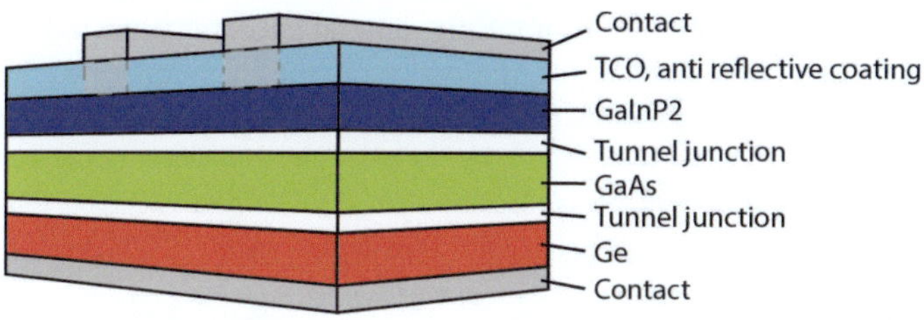

Fig. 8.43 Schematic layer sequence of a triple-junction solar cell. Mono-crystalline Ge is used for the bottom cell, followed by a GaAs middle cell and a GaInP2 top cell. Each cell includes a p/n-junction. The three cells are internally electrically connected by two tunnel junctions

levels have to be high enough to reduce the corresponding space charge layer widths sufficiently to allow effective carrier tunnelling through the junction (Fig. 8.42).

The increased efficiency of a tandem solar cell compared to a single-junction cell can be further improved by combining more sub-cells in series, thus further reducing thermalization losses for a wider range of the solar spectrum. So far, multijunction cells with up to six sub-cells have been realized. As an example, we show in Fig. 8.43 the schematic layer sequence of a triple-junction solar cell. Such cells have reached efficiencies of up to 40% in the laboratory, but require expensive deposition methods such as MOCVD or MBE for the realization of defect-free epitaxial growth of the different sublayers.

8.6.3 Impurity-Band Photovoltaics (Optical Up-Conversion)

Similar to optical down-conversion, it is in principle possible to achieve an optical excitation in the absorber by combining two low energy photons with energies smaller than E_{gap}. At the relatively low intensity levels of unconcentrated sunlight, this usually requires the presence of a defect band in the band gap of the absorber to serve as an intermediate electronic state. However, optical up-conversion competes directly with the very efficient recombination of excited carriers via the same intermediate defect band. Thus, no convincing examples of up-conversion in real solar cells have been reported so far.

8.6.4 Impact Ionization (Carrier Multiplication)

This process corresponds to the inverse process of Auger recombination: an optically excited hot electron with $E_{kin} > E_{gap}$ creates an additional electron-hole pair by Coulomb interaction with other carriers. For photons with energy $>2E_{gap}$, this would result in a quantum efficiency for the generation of electron-hole pairs larger than one. Indeed, carrier multiplication has been reported for optical absorption in semiconductor quantum dots, where due to the small volume Coulomb interaction between carriers is very strong. To make use of carrier multiplication in photovoltaics, however, the extraction of multiply generated carriers has to be faster than the competing Auger recombination. Again, no convincing evidence for a significant improvement of solar cell efficiencies by this process has been demonstrated so far (Fig. 8.44).

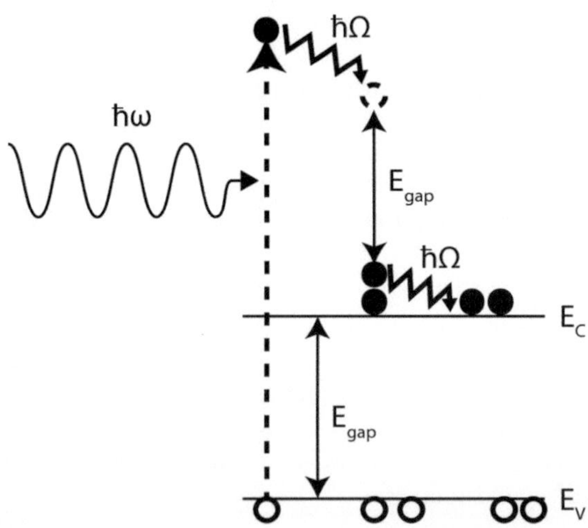

Fig. 8.44 Multiple generation of optically excited carriers via the absorption of high-energy photons followed by impact ionization

8.7 Energy Payback Times of Solar Cells

An important aspect that has been and needs to be discussed and considered in view of a large scale global use of solar cells (as well as for all other renewable energy technologies) for a sustainable future energy supply is their energy payback time and the related lifetime energy harvesting factor. This conceptually very simple question turns out to be very complicated and controversial when applied to real products. For solar cells, one has to take into account on the one side the energy costs for mining raw materials such as Cu, Si, Ga or In, the energy spent for the construction of production facilities, the energy required for the solar cell production itself, and the energetic deployment and recycling costs. On the other side, the energy harvested by a solar cell during its lifetime needs to be estimated, which depends on the solar cell efficiency, their placement in the environment, their lifetime, maintenance conditions, etc.

As of today, most solar cell technologies are estimated to have energy payback times between one and a few years and energy harvesting factors around 10. However, the current use of photovoltaics is less driven by physical considerations than much more by economic boundary conditions, involving the costs and long-term environmental aspects of competing energy sources and the resources needed for solar cell production. Therefore, we refrain from a further discussion of this delicate and indeed very complicated aspect.

Further Reading

- Partially commercial solar irradiance and weather data service by Solargis: solargis.com. Accessed 29 Aug 2022
- Development of solar cell efficiency: www.nrel.gov/pv/cell-efficiency.html. Accessed 28 Aug 2022
- Sugathan, V., John, E., Sudhakar, K.: Recent improvements in dye sensitized solar cells: a review. Renew. Sustain. Energy Rev. **52**, 54–64 (2015)
- Wang, R., et al.: A review of perovskites solar cell stability. Adv. Funct. Mater. **29**, 1808843 (2019)

References

1. BP: Statistical Review of World Energy, 70 (2021)
2. Poortmans, J., Arkhipow, V. (eds.): Thin Film Solar Cells. Wiley, Chichester (2006)
3. Chapin, D.M., Fuller, C.S., Pearson, G.L.: A new silicon p-n junction photocell for converting solar radiation into electrical power. J. Appl. Phys. **25**, 676 (1954)
4. Schröder, D.K.: Carrier lifetimes in silicon. IEEE Trans. Electron Devices **44**(1), 160–170 (1997)
5. Shockley, W., Queisser, H.J.: Detailed balance limit of efficiency of p-n junction solar cells. J. Appl. Phys. **32**, 510 (1961)
6. Koynov, S., et al.: Appl. Phys. Lett. **88**, 203107 (2006)

Thermoelectrics

9

Abstract

The last chapter deals with the field of thermoelectrics. As the physical basis of these devices the Seebeck and Peltier effect are explained as well as the implementation in thermoelements and thermoelectric generators (TEGs). The requirements of efficient thermoelectric conversion are presented together with state-of-the-art materials for these. The chapter concludes with the possible fields of application.

9.1 Basic Physics of Thermoelectricity

Thermoelectricity is a physical process allowing the direct transformation of heat into electricity without any moving parts. Therefore, it is highly reliable and basically maintenance-free, so that thermoelectric generators were used early on for deep space missions such as the Voyager spacecrafts launched in 1977 [1]. Their supply with electrical energy is based on thermoelectricity using radioactive heat sources. Since waste heat is a by-product of many energy-conversion processes, thermoelectric converters could have a wide range of applications also in the context of renewable energy. The driving force behind thermoelectricity is a heat flow caused by a thermal gradient in a conductive material. The physical basis for this is the *Seebeck effect* (also referred to as *thermopower*) in electrical conductors such as metals and doped semiconductors. Assume a conductive bar of length d extending in x-direction, with two different temperatures T_1 and T_2 at both ends (cf. Fig. 9.1). If a volt-meter is attached to both ends, it will measure a thermo-voltage U_{th} caused by different electrical potentials ϕ_1 and ϕ_2 at the cold and hot end. The Seebeck effect now states that the gradient of the electrical potential for zero electrical current through the bar is proportional to the temperature gradient across the bar, with a proportionality factor α, the so-called *Seebeck coefficient*.

$$\mathcal{E} = -\text{grad}\,\phi = \alpha\,\text{grad}\,T$$

M. Stutzmann and C. Csoklich, *The Physics of Renewable Energy*, Graduate Texts in Physics,
https://doi.org/10.1007/978-3-031-17724-8_9

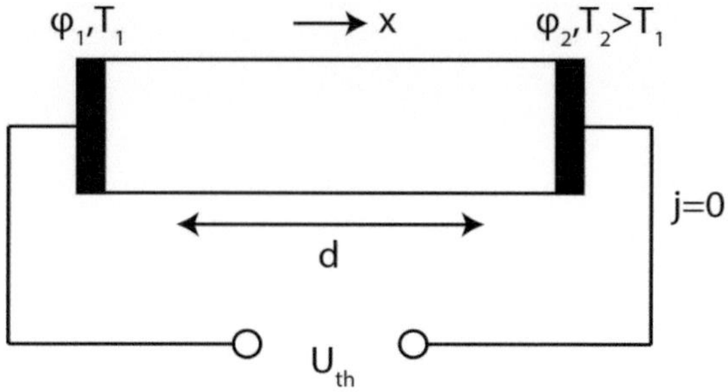

Fig. 9.1 The Seebeck effect for a linear conducting bar. $T_2 > T_1$ are the absolute temperatures of the hot and cold ends, giving rise to the thermo-voltage U_{th} for zero current flow, $j = 0$

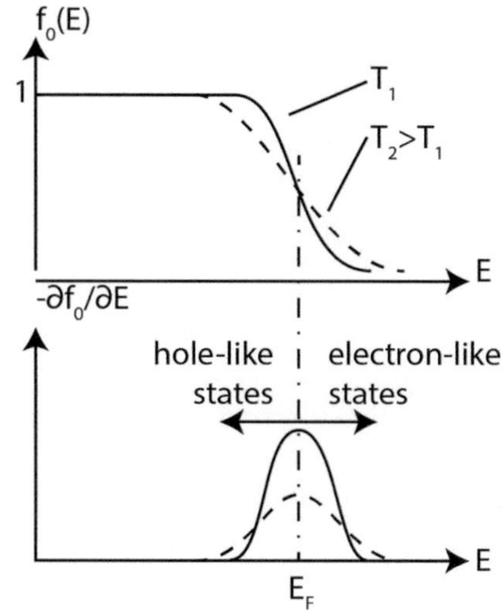

Fig. 9.2 Schematic sketch of the equilibrium Fermi-function f_0 for two different temperatures T_1 and $T_2 > T_1$. The negative derivative of f_0 with respect to the energy E visualizes the thermal excitation of electrons around the Fermi energy E_F compared to $T = 0$. Below E_F, electrons are missing and form a distribution of thermally excited hole-like states, whereas excited electron-like states appear above E_F with increasing temperature

In the simple 1D-case shown in Fig. 9.1 above and for homogeneous materials, this can also be approximated by the macroscopic relations:

$$\text{grad}\, T = \frac{T_2 - T_1}{d}; \qquad U_{th} = \phi_2 - \phi_1$$

$$\boxed{U_{th} = \alpha(T_2 - T_1)} \qquad \text{for j=0; } [\alpha]\text{=V/K} \tag{9.1.1}$$

The emergence of a thermopower can be derived from the differences of the electronic Fermi-function at the hot and cold end, as sketched in Fig. 9.2.

$$f_0(E) = \left(\exp\left(\frac{E - E_F}{kT}\right) + 1\right)^{-1}$$

To understand how differences in the distribution of hole- and electron-like states lead to the Seebeck effect, we note that

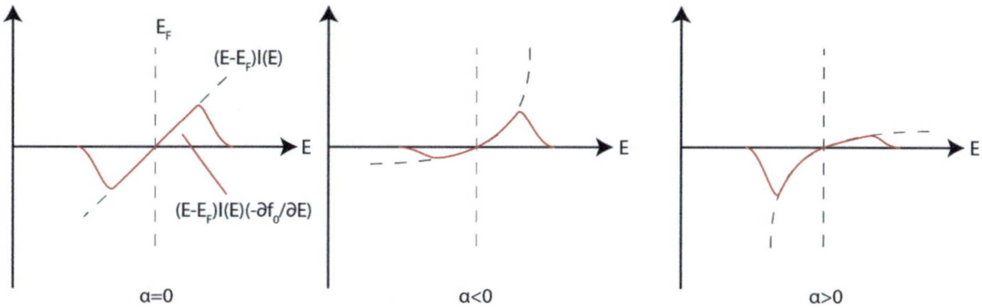

Fig. 9.3 Schematic drawing of symmetric and asymmetric distributions of electron- and hole-like states in the vicinity of the Fermi energy of a metal. The energy $(E - E_F) \cdot l(E)$ transported per electron or hole is indicated by the dashed black lines, the red curves indicate a symmetric ($\alpha = 0$, left), an asymmetric, electron-like ($\alpha < 0$, middle), and an asymmetric, hole-like character ($\alpha > 0$, right)

- Thermally excited electrons around E_F have an excess kinetic energy of $\approx E - E_F$
- For a finite Seebeck coefficient $\alpha \neq 0$, an asymmetry between electron-like and hole-like states around E_F is necessary.
  - In semiconductors, this can be easily established by doping, giving rise to a large difference in the equilibrium densities of electrons and holes for a given temperature.
  - In metals, more subtle differences in the densities of state, the effective masses or the mean free paths $l(E)$ for electron- and hole-like states are responsible for non-zero Seebeck coefficients. For example, the characteristic transported thermal energy per electron or hole is proportional to $(E - E_F) \cdot l(E)$ (cf. Fig. 9.3).

A simplified qualitative argument for the appearance of a thermo-voltage under the presence of a thermal gradient is that in electron-like metals electrons at the hot end have a higher kinetic energy due to a higher thermal velocity than electrons at the cold end. Thus, there is an overall net movement of electrons from the hot to the cold end. As a consequence the cold end is charged negatively, while the hot end is charged positively. The resulting electric field counteracts the further diffusion of electrons from hot to cold, until a steady state with a constant thermo-voltage U_{th} is established.

The situation in an n-type semiconductor is described in Fig. 9.4. Due to n-type doping, the Fermi level is closer to the conduction than to the valence band edge, making electrons the majority carriers. At the cold end, the density n of thermally excited electrons in the conduction band is smaller than at the hot end. This difference in electron density again causes a diffusion current of electrons from hot to cold. The resulting electrical field drives a counteracting drift current of electrons from cold to hot. Steady state and a constant thermo-voltage are reached, when drift and diffusion current cancel each other. An analogous picture holds for p-type semiconductors with thermally excited holes.

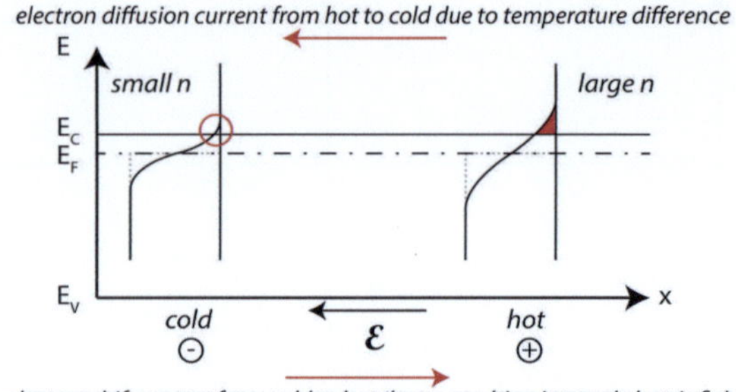

Fig. 9.4 N-type semiconductor with a thermal gradient. Sketched are the different Fermi distributions and the resulting electron densities in the conduction band. See text for further details

For both, metals and semiconductors, the dominant type of carriers (electrons or holes) present is indicated by the sign of the thermo-voltage at the cold end. This can be used to quickly determine the doping type of a semiconductor. Furthermore, the full steady-state thermo-voltage is obtained when drift and diffusion currents cancel each other exactly, i.e. for zero internal current density, $j = 0$.

For $j = 0$ the drift current generated by the internal electric field is opposite and equal to the diffusion current generated by ΔT.

Thermocouples: To measure the Seebeck coefficient α of a sample of material B, one needs electrical contacts usually consisting of a different material A. As indicated below this also generates a thermo-voltage. To avoid further complications due to additional materials in the voltmeter, the latter should be maintained at a constant temperature T_0.

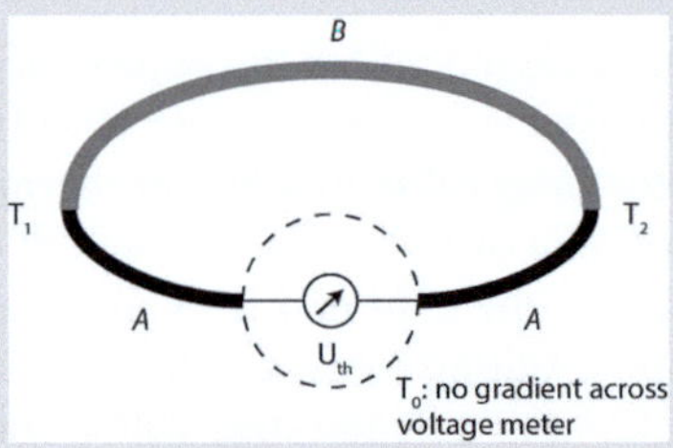

For the thermo-couple arrangement shown above, the expected thermopower can be calculated using the Seebeck relation by integration over the different segments of the closed conductor loop:

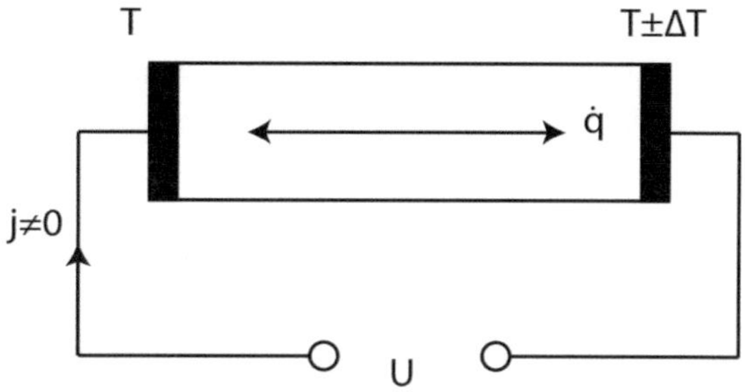

Fig. 9.5 The Peltier effect. See text for details

$$
\begin{aligned}
U_{th} &= -\oint \mathcal{E} \cdot \mathbf{ds} = -\oint \alpha \,\mathrm{grad}T \cdot \mathbf{ds} = -\int_{T_0}^{T_1} \alpha_A \mathrm{d}T - \int_{T_1}^{T_2} \alpha_B \mathrm{d}T - \int_{T_2}^{T_0} \alpha_A \mathrm{d}T \\
&= \int_{T_1}^{T_2} (\alpha_A - \alpha_B)\,\mathrm{d}T \\
&\approx (\alpha_A - \alpha_B)(T_2 - T_1) \qquad \text{for small } \Delta T
\end{aligned}
$$

Therefore, with such a thermo-couple the Seebeck coefficient of test material B can be calculated from the applied temperatures relative to the Seebeck coefficient of a reference material A. Here, one has to further consider the following aspects:

- In general $\alpha = \alpha(T)$ is strongly temperature dependent.
- For the reference material, a highly conductive metal with a small Seebeck coefficient such as Pt should be used.
- For non-zero current density $j \neq 0$, power can be extracted from a thermo-couple (cf. (9.2.1)). This also reduces the thermo-voltage U_{th}.

Also the inverse of the thermopower can be observed and is known as the *Peltier effect*: here a current density j due to an externally applied voltage U creates a heat flow density $\dot{q}$ which in turn generates a temperature gradient between the electrical contacts (Fig. 9.5). The direction of the temperature gradient depends on whether the conductivity in the Peltier element is electron- or hole-like. The Peltier coefficient $\tilde{\pi} = \alpha T$ is directly related to the Seebeck coefficient α and the absolute temperature T.

A quantitative treatment of the Seebeck and Peltier effect can be performed based on the Boltzmann equation for the Fermi-distribution $f(\mathbf{r}, \mathbf{k}, t)$ in phase space:

$$
\left.\frac{\partial f(\mathbf{r}, \mathbf{k})}{\partial t}\right|_{scattering} = \frac{df}{dt} = \underbrace{\frac{\partial f}{\partial t}}_{0 \text{ in stationary state}} + \underbrace{\dot{\mathbf{k}}\, \mathrm{grad}_{\mathbf{k}} f}_{\frac{\partial f}{\partial k}\frac{\partial k}{\partial t}} + \underbrace{\dot{\mathbf{r}}\, \mathrm{grad}_{\mathbf{r}} f}_{\frac{\partial f}{\partial r}\frac{\partial r}{\partial t}}
$$

Here the last term on the right side allows the inclusion of spatial gradients of temperature and their effect on the Fermi distribution f. Solving this equation assuming different scattering mechanisms gives the following formulae for the Seebeck-coefficient and its T-dependence in metals and semiconductors (with typical orders of magnitude indicated at the right side):

$$\boxed{\alpha_{met} = -\frac{\pi^2}{2}\frac{k_B}{e}\frac{k_B T}{E_F}}$$ for electron-like metals ($\approx \mu$V/K)

$$\boxed{\alpha_n = -\frac{k_B}{e}\left[\frac{E_C - E_F}{k_B T} + \frac{5}{2} + r\right]}$$ for n-type semiconductors ($\approx$ m V/K)

$$\boxed{\alpha_n = +\frac{k_B}{e}\left[\frac{E_F - E_V}{k_B T} + \frac{5}{2} + r\right]}$$ for p-type semiconductors ($\approx$ m V/K

Here, r is a scattering factor which depends on the dominant scattering mechanism of the majority carriers (e.g. $r = -1/2$ for deformation potential scattering by acoustic phonons or $r = 3/2$ for scattering at ionized impurities).

9.2 Thermoelectric Generators (TEGs)

Thermoelectric generators are devices to harvest electrical energy with the help of suitably assembled thermoelectric elements. As shown in Fig. 9.6, they are based on pairs of n- and p-type semiconductors (called n- and p-"legs") connected parallel and/or in series to larger assemblies, depending on the desired output voltage or current.

Obvious requirements for suitable semiconductor materials for the legs are:

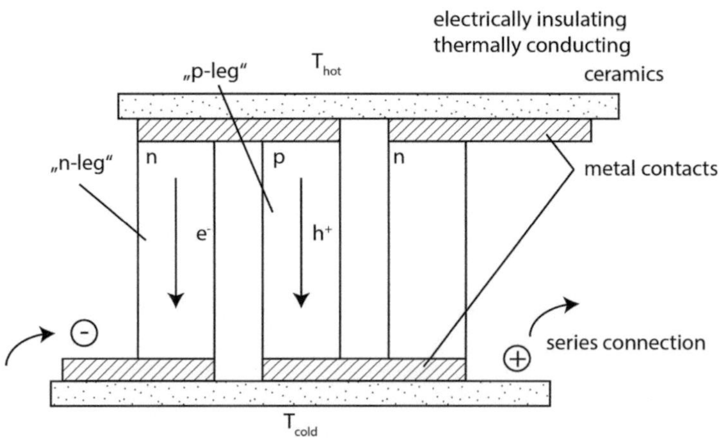

Fig. 9.6 Schematic layout of a thermoelectric generator. "Legs" of alternating n- and p-type semiconducting bars are connected in series by ohmic metal contacts. The electrical structure is encased by a suitable ceramic to make contact to the hot and cold heat reservoirs, setting up a constant thermal gradient through the legs. Arrows indicate the diffusion direction of the majority carriers in the legs

- a large Seebeck-coefficient α ($[\alpha]$=V K^{-1})
- a high electrical conductivity $\sigma = en\mu$ ($[\sigma]$=A $V^{-1}m^{-1}$)
- a low thermal conductivity κ ($[\kappa]$= W $K^{-1}m^{-1}$)
- sufficient structural and electrical stability in the temperature range applied.

To compare the suitability of different thermoelectric materials one uses the *thermoelectric figure of merit* ZT:

$$\boxed{Z \cdot T \equiv \frac{\alpha^2 \sigma}{\kappa} \cdot T} \qquad ([Z] = \mathrm{K}^{-1};\ [Z \cdot T] = 1) \tag{9.2.1}$$

This figure of merit determines the maximum theoretical efficiency of a TEG:

$$\boxed{\eta_{max} = \underbrace{\frac{T_H - T_C}{T_H}}_{\eta_{Carnot}} \cdot \frac{\sqrt{1 + ZT_{ave}} - 1}{\sqrt{1 + ZT_{ave}} + \frac{T_C}{T_H}}} \tag{9.2.2}$$

with the average temperature $T_{ave} = \frac{1}{2}(T_H + T_C)$. T_H and T_C are the absolute temperatures of the hot and cold reservoirs. Note that:

$$\eta_{max} \rightarrow \eta_{Carnot} \qquad \text{for} \qquad Z \cdot T \rightarrow \infty$$

The formula above is only valid for $T_H - T_C << T_C$. Otherwise, since $Z = Z(T)$ is strongly dependent on temperature, one has to obtain η_{max} by integration:

$$\eta_{max} = \frac{T_H - T_C}{T_H} \frac{1}{T_H - T_C} \int_{T_C}^{T_H} \frac{\sqrt{1 + Z(T) \cdot T} - 1}{\sqrt{1 + Z(T) \cdot T} + \frac{T_C}{T_H}} dT$$

For energy conversion in TEGs, only materials with $ZT \geq 1$ are commercially viable. This can be seen from the maximum theoretical efficiencies plotted in Fig. 9.7 and listed in Table 9.1.

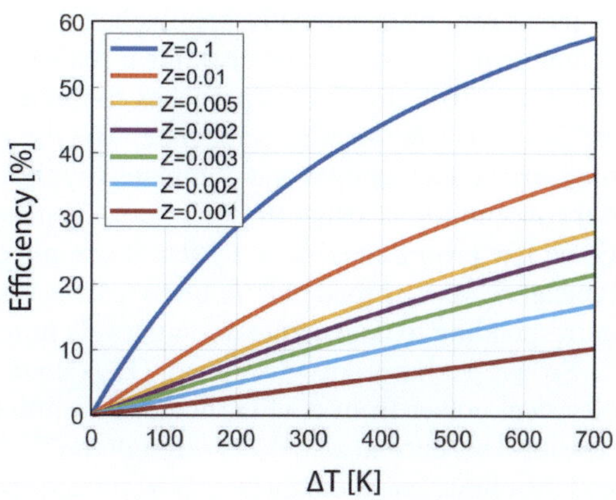

Fig. 9.7 Dependence of the theoretical TEG efficiency on the figure of merit and the temperature difference between hot and cold heat reservoirs. The cold reservoir was assumed to be at room temperature, 300 K

Table 9.1 Theoretical efficiencies of TEGs for different T_H and $T_C = 300\,\text{K}$ in comparison to the Carnot efficiency

T_H	η_{Carnot}	η_{max} $(ZT = 0.1)$	η_{max} $(ZT = 1)$	η_{max} $(ZT = 10)$
500 K	0.4	0.01	0.08	0.24
1000 K	0.7	0.03	0.17	0.45

In addition to T, $Z = (\alpha^2\sigma)/\kappa$ also depends strongly on the carrier densities n or p. In particular, we can approximate for n-type materials:

$$\alpha \propto \frac{k_B T}{E_F} \quad \text{with} \quad E_F = \frac{\hbar^2}{2m^*}(3\pi^2 n)^{2/3} \qquad \text{in metals}$$

$$\alpha \propto \frac{E_C - E_F}{k_B T} = \ln\left(\frac{N_C}{n}\right) \qquad \text{in semiconductors (n-type)}$$

$$\sigma = en\mu_n$$

$$\kappa = \kappa_{phonon} + \kappa_{electron} \quad \text{with} \quad \kappa_{electron} = L\sigma T$$

The last equation is the *Wiedemann-Franz law* in metals, with the *Lorentz factor L* for free electrons:

$$L = \frac{\pi^2}{3}\left(\frac{k_B}{e}\right)^2 = 2.4 \times 10^{-8}\text{J}^2\text{K}^{-2}\text{C}^{-2}$$

Finally, the phonon contribution to the thermal conductivity is given by:

$$\kappa_{Phonon} \propto T^3 \text{ for } T < \Theta_D; \qquad \kappa_{Phonon} \propto 1/T \text{ for } T \geq \Theta_D \tag{9.2.3}$$

with Θ_D the Debye-temperature for acoustic phonons.

A qualitative view of the dependence of the different contributions to the quantity $Z = (\alpha^2\sigma)/\kappa$ on charge density n in n-type semiconductors is shown in Fig. 9.8. The Seebeck coefficient α decreases logarithmically with the carrier concentration, as the Fermi-level gradually approaches the conduction band edge. The electrical conductivity σ is negligible on a linear scale for small carrier concentrations. Above about $n = 10^{20}\,\text{cm}^{-3}$, σ approaches the minimum metallic conductivity of about $10^4\,\Omega^{-1}\,\text{cm}^{-1}$ in the metallic regime. The thermal conductivity κ is independent of the carrier concentration and dominated by the phonon-contribution κ_{phonon} for carrier concentrations below $10^{20}\,\text{cm}^{-3}$ and then increases according to the Wiedemann-Franz law proportional to σ by the electronic contribution $\kappa_{electron}$. As a result, Z typically shows a broad maximum for carrier concentrations around $n = 10^{20}\,\text{cm}^{-3}$, corresponding to degenerate doping levels in semiconductors.

So far, we have only discussed TEGs under open-circuit conditions. To extract electrical power from a TEG, however, a finite current has to flow through a load resistor. This current also flows through the TEG which, due to the highly doped legs and the ohmic metal contacts, can be treated like an internal resistor, R_{TEG}. This is

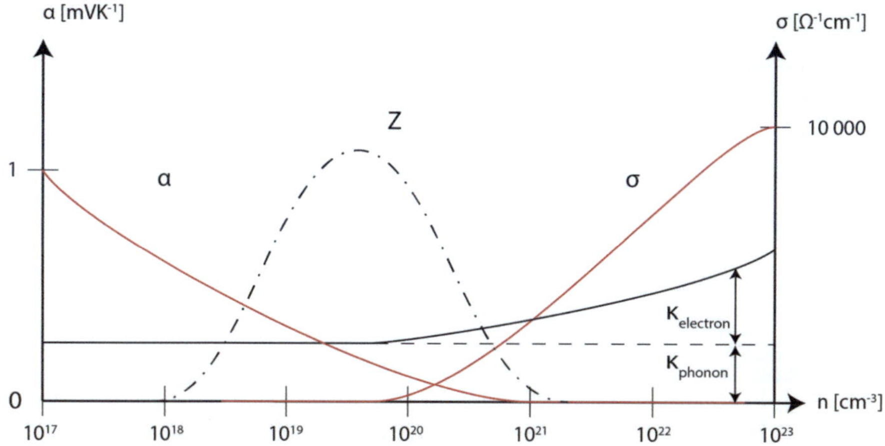

Fig. 9.8 Qualitative semi-logarithmic view of the different contributions to $Z = (\alpha^2\sigma)/\kappa$ as a function of the carrier density n in an n-type semiconductor. See text for details. Since the thermal conductivity is basically constant, a similar broad maximum around $n = 10^{20}\,\mathrm{cm}^{-3}$ as for Z is also obtained for its numerator, $\alpha^2\sigma$, which is called the *power factor*

shown in Fig. 9.10. For a TEG with N p/n-legs in series and an effective Seebeck coefficient α operated with a temperature difference ΔT, the open-circuit voltage is (Fig. 9.9):

$$U_{oc} = N \cdot \alpha \cdot \Delta T$$

The voltage U dropping across the load resistor is then:

$$U = R_{load} \cdot I = U_{oc} - R_{TEG} \cdot I$$

The maximum output power is reached for $R_{TEG} = R_{load}$:

$$P_{max} = U_m \cdot I_m = \frac{U_{oc}^2}{4R_{TEG}}$$

To quantify the performance of TEGs, in addition to the current-voltage characteristics a direct measurement of the figure of merit ZT can be achieved by the so-called *Harman method*:

1. A current I is passed through the TEG and produces via the Peltier effect a heat flow $\dot{Q} = \tilde{\pi} \cdot I = \alpha \cdot T \cdot I$
2. In steady state this causes a temperature difference ΔT, given by $\dot{Q} = \alpha T I = \kappa \Delta T$.
3. The voltage U_{stat} across the TEG in steady state is measured: ($R_{TEG} \propto 1/\sigma$):

$$U_{stat} = U_{oc} - R_{TEG} I$$

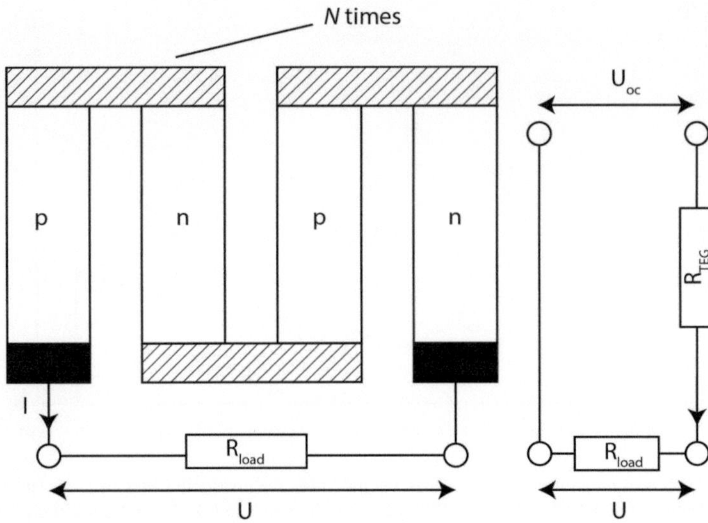

Fig. 9.9 Basic structure of a thermoelectric generator with N p/n-legs in series and a load resistor R_{load}. The corresponding equivalent circuit is shown on the right

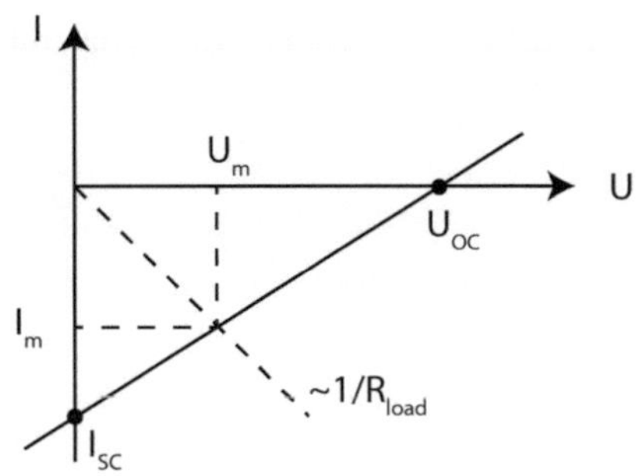

Fig. 9.10 Current-voltage characteristics of a TEG. Since the TEG acts like a resistor, this leads to a linear characteristics with slope R_{TEG}^{-1}. The corresponding fill factor is much smaller compared to the fill factor of solar cells with diode characteristics

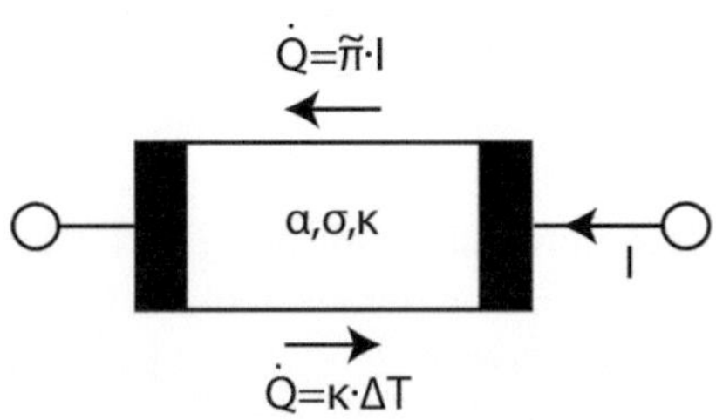

Fig. 9.11 Illustration of the Harman method

4. The current I is stopped abruptly and immediately after $U_{oc} = \alpha \Delta T$ is measured.
5. ZT can then be calculated from U_{oc} and U_{stat} (Fig. 9.11).

$$ZT = \frac{\alpha^2 \sigma}{\kappa} T = \frac{U_{oc}}{U_{stat} - U_{oc}} \tag{9.2.4}$$

Table 9.2 Applications for TEGs

With fuel heating	With waste heat
Deep space missions-RTG	Waste burning facilities
Pace makers	Waste heat in cars
Light houses, transmitter stations	Geothermal plants
Corrosion protection	Miniature converters (watches, sensors,..)
Mobile power supplies	Solar climate regulation (venting)
Electrical backup systems for solar plants	Space missions near the sun

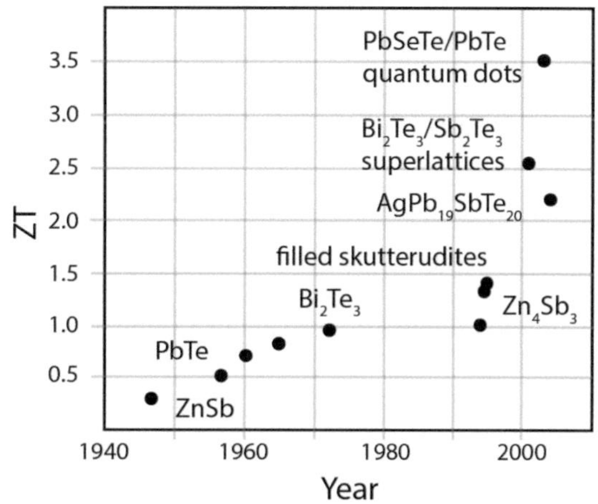

Fig. 9.12 Historical development of the figure of merit ZT for different materials and nanostructures

Some additional remarks concerning TEGs:

- For $T_H \geq 1200\text{K}$, thermo-photovoltaics becomes more efficient than thermoelectrics [2]. For $T_H \geq 2500\text{K}$, photovoltaic energy conversion is preferable.
- The efficiencies of current commercial TEGs are in the range of 1–10% [3].
- Some examples for applications of TEGs are listed in Table 9.2.

Current research to develop systems with high ZT values is focussed on complex compound materials and the specific use of nanostructures. The aim is to decrease the thermal conductivity of thermoelectric systems, while keeping the Seebeck coefficient and the electrical conductivity high or even increasing them. A detailed treatment of different approaches would go beyond the scope of this book, so we just mention a few headlines (Fig. 9.12).

Early TEGs were based on crystalline SiGe alloys, since these could be produced as large crystals with good quality and could be equally well doped n- and p-type for the different TEG legs. SiGe reached $ZT = 1$ for a temperature of 1200 K and was used primarily in early deep space missions. Most other thermoelectric materials also used compound semiconductors with heavy atoms such as PbTe or Bi2Te3, because

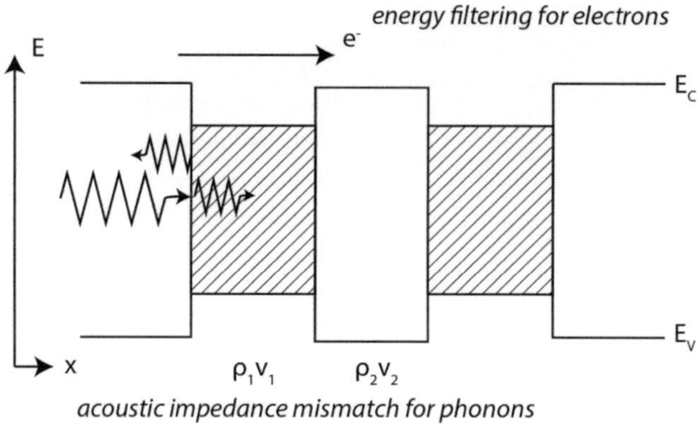

Fig. 9.13 Energy band diagram of a semiconductor superlattice. The energetic filtering of electrons or holes by the periodic band gap modulation plus the multiple scattering of phonons by the acoustic impedance mismatch between the two constituent materials help to increase ZT for such structures

they combined a low thermal conductivity due to alloy scattering and the heavy atoms involved with reasonable Seebeck coefficients and electrical conductivities. Further advances were made around 1990–2000 with novel materials beyond binary alloys, especially chlathrites and skudderudites featuring even lower thermal conductivities due to their large crystal unit cells and so-called "rattling atoms", i.e. heavy atoms vibrating with relatively low energies yet large amplitudes, thereby providing efficient scattering centers for other lattice phonons ("phonon glass-electron crystal" Behaviour). With this approach ZT-values exceeding 2 were realized, e.g. for the compound AgPb19SbTe20. Even higher ZT-values exceeding 2.5 were reached for Bi2Te3/Sb2Te3 superlattices. As sketched in Fig. 9.13, such superlattices combined additional multiple phonon scattering at the layer interfaces due to a sudden change in the acoustic impedance ρv_{sound} (density times sound velocity) of the different materials, and the energy filtering of electrons passing through the superlattice conduction band. A negative aspect of the use of such compound superlattices are their high production costs.

Even higher ZT-values above 3 were reported for thermoelectric materials based on compound semiconductor nanostructures such as quantum dots or quantum wires. Here, improvements of ZT have been related to an increased surface or interface scattering of phonons with bulk mean free paths exceeding the nanostructure size. In addition, positive effects of the altered electronic density of states in 1D- or 0D-confined systems on the Seebeck coefficient have been postulated.

Further Reading

- Rowe, D.M. (ed.): Thermoelectrics Handbook: Macro to Nano. CRC Press, Boca Raton (2017).
- Radioisotope TEG with plutonium-238 on the Cassini and other space probes: solarsystem.nasa.gov/missions/cassini. Accessed from 29 July 2022.

- Wei, Jiangtao, et al.: Review of current high-ZT thermoelectric materials. J. Mat. Sc. **55**, 12642–12704 (2020).

References

1. National Aeronautics and Space Administration, Voyager spacecraft website: https://voyager.jpl.nasa.gov/mission/spacecraft/. Accessed from 31 July 2022
2. LaPotin, A., et al.: Thermophotovoltaic efficiency of 40%. Nature **604**, 288 (2022)
3. Champier, Daniel: Thermoelectric generators: a review of applications. Energy Convers. Manag. **140**, 167–181 (2017)

Knowledge Check

Problems of Chap. 1

1.1 What is the unit of energy expressed in terms of the SI units kg, m, and s?

1.2 What is the meaning of the abbreviation "BTU"?

1.3 Show that the kinetic energy of a moving mass m is $1/2 \cdot mv^2$ in classical mechanics.

1.4 According to the Noether theorem, which symmetry leads to energy conservation?

1.5 What is the basic metabolic rate of an average human being in rest?

(A) 30 W
(B) 80 W
(C) 150 W
(D) 280 W

Problems of Chap. 2

2.1 Provide the formula for the energy density E/V of a plate capacitor with capacitance C charged up to a voltage U.

2.2 Which of the following forms of energy storage has the lowest volumetric energy density E/V?

(A) a commercially available supercapacitor
(B) a Li-ion battery

M. Stutzmann and C. Csoklich, *The Physics of Renewable Energy*,
Graduate Texts in Physics,
https://doi.org/10.1007/978-3-031-17724-8

(C) water with a height difference of 100 m
(D) a high temperature phase change salt

2.3 According to the Nernst equation, how does the voltage U developing at a solid electrode immersed in a liquid electrolyte depend on the concentration c of dissolved electrode ions?

2.4 Give the overall reaction equation for the nuclear fission of ^{235}U.

Problems of Chap. 3

3.1 What is the total power emitted by an ideal black body with a temperature 1000 K and a surface area of 1 m^2? (Stefan–Boltzmann constant: $\sigma = 5.67 \times 10^{-8}$ in SI units)

3.2 What is the overall fusion reaction in our sun, considering all conservation laws?

3.3 What is the effective surface temperature of the Sun? Describe two ways how this temperature can be determined here on Earth.

3.4 What would be the average surface temperature of the Earth without the greenhouse effect?

1. Give the equation based on the balance between absorbed and emitted energy using an average reflectivity of 30% for the incoming radiation.
2. Give the resulting value for the average surface temperature.

3.5 What is the current average value of the Solar Constant (SC)? How large (in %) are its seasonal changes?

(A) $\pm 1\%$
(B) $\pm 5\%$
(C) $\pm 10\%$

3.6 If all of the radiation of the Sun impinging on the Earth could be captured and used, how much time of sunshine would be necessary to satisfy the current energy need of mankind?

(A) 1 min
(B) 1 h
(C) 1 day
(D) 1 week

Problems of Chap. 4

4.1 What is the dispersion relation v(λ) of deep water waves? Is there a difference to the relation for shallow water waves?

4.2 Give the formula for the gradient of the gravitational acceleration caused by the sun on the two sides of the earth pointing towards and away from the sun.

4.3 What is the main qualitative difference between solar and lunar tides when calculating gradients of gravitational and centrifugal forces?

4.4 Why can local tidal waves be much higher than estimated under the assumption of a global ocean?

Problems of Chap. 5

5.1 Why is there a low pressure region at the intertropical conversion zone (ITCZ)?

5.2 What is the areal power density P/A of wind with velocity v?

5.3 Draw a diagram of the variation of pressure $p(x)$ when air passes through the rotor plane of a wind engine at $x = 0$ (the pressure far away from the wind engine is p_0).

5.4 Draw the Betz efficiency of a wind engine as a suitable function of the wind speeds before (v_1) and after the engine (v_2) and give the limiting values (maximum efficiency, efficiency for $v_2 = 0$ and point of zero efficiency).

5.5 Draw the vector diagram defining the velocity of attack v_A of a lift type rotor.

5.6 What is the tip speed ratio λ_s?

5.7 A turbine wing has the glide number $G = 100$ and experiences a drag force of $F_W = 1000\,\text{N}$. What is the corresponding lift force F_A?

5.8 What are the main loss mechanisms of a lift type 3-blade rotor at

(a) very low tip speed ratio λ_s?
(b) very high tip speed ratio λ_s?

5.9 Wind with a velocity of $50\,\text{km}\,\text{h}^{-1}$ flows through an area A. What is the corresponding power density P/A, using a density of $\rho = 1.2\,\text{kg}\,\text{m}^{-3}$ for the flowing air?

5.10 Wind with the velocity $v_1 = 50\,\mathrm{km\,h^{-1}}$ enters an ideal wind engine with an area of $A = 10000\,\mathrm{m^2}$ and exits with a velocity of $25\,\mathrm{km\,h^{-1}}$.

1. What is the Betz efficiency of this wind engine?
2. How does this efficiency change when the area of the wind engine is reduced to $1000\,\mathrm{m^2}$?

5.11 A blade segment has drag coefficient $c_W = 0.3$, a perpendicular area $A_\perp = 1\,\mathrm{m^2}$, a lift coefficient $c_A = 0.9$ and a parallel area $A_{||} = 10\,\mathrm{m^2}$. The velocity of attack is $v_A = 10\mathrm{m\,s^{-1}}$. What is the glide number G of this segment?

5.12 What is the realistic maximum efficiency of a modern 3-blade wind engine? At approximately which tip speed ratio λ_S is this efficiency achieved?

5.13 Give the empirical formula for the average wind velocity as a function of height h above ground.

5.14 How does the breadth $b(r)$ parallel to the velocity of attack of an optimized rotor blade scale with the tip speed ratio λ_S and the distance r from the center of the wind engine?

Problems of Chap. 6

6.1 How large is the average geothermal gradient in the Earth's crust?

6.2 At a given high enthalpy site the local heat flux to the surface is $300\,\mathrm{mW\,m^{-2}}$ and the local thermal conductivity is $3\,\mathrm{W/(Km)}$. How large is then the thermal gradient driving this heat flux?

6.3 What is the definition of the efficiency η_{HP} of a heat pump and what is its thermodynamic limit?

6.4 A heat pump operates between the temperature levels $100\,°\mathrm{C}$ and $25\,°\mathrm{C}$. What is the maximum efficiency of this heat pump?

6.5 Sketch the optimized wavelength-dependence of the absorbance for a non-concentrating solar-thermal panel.

6.6 What is the maximum achievable concentration factor C_{max} of direct sunlight using a linear concentrator?

6.7 A solar tower is illuminated by a heliostat field with a concentration factor of $C = 1000$. What is the maximum achievable absorber temperature?

Problems of Chap. 7

7.1 What is the typical carbon fixation capacity of so-called C4 plants in mg CO_2 per 100 cm^2 leave area and hour?

(A) 10 mg
(B) 30 mg
(C) 70 mg
(D) 120 mg

7.2 What is the current overall efficiency of global photosynthesis (energy stored in annually produced biomass vs. integrated incoming solar irradiation)?

(A) 0.01%
(B) 0.1 %
(C) 1%
(D) >1%

7.3 What are the two main intermediate energy storage systems in photosynthesis and to which photosystem are they linked?

7.4 Sketch the primary and secondary reactions of photosynthesis with the input and output molecules and the intermediate energy storage compounds.

7.5 Describe the role of C-C double bonds in chlorophyll for the absorption of light.

7.6 How many chlorophyll molecules are typically present in the antenna complexes of the photosystems?

7.7 1. What is the distance dependence of the Förster transfer rate between neighboring chlorophyll molecules in an antenna complex?
2. What does this distance dependence say about the type of interaction between the molecules?

7.8 What is synthesis gas or "Syn-gas"?

Problems of Chap. 8

8.1 Draw the "Feynman diagram" of photovoltaics.

8.2 The absorber layer of a solar cell has an absorption coefficient of $\alpha = 10^4\,cm^{-1}$ at a wavelength of 600 nm. What is the required absorber thickness d to absorb 95% of the incoming photons of that wavelength?

8.3 Describe the time derivative of the minority carrier density n in a p-type absorber by the recombination and generation rates R and G, respectively.

8.4 For the case of bimolecular recombination, what is the dependence of the photo-generated excess carrier density Δn on the generation rate G?

8.5 Sketch the energy band diagram of a Si diffusion cell including the BSF and the built-in voltage U_{bi}.

8.6 Explain why the efficiency of a solar cell goes to zero in the limit of small and of large band gaps of the absorber layer.

8.7 Which combination of band gaps in tandem solar cells is optimum for $AM = 1.5$ illumination conditions?

8.8 What is the time and spatial dependence $E(x, t)$ of the electric field of a photon with frequency ν in a matter with complex index of refraction $(n + i\kappa)$?

8.9 What is the average generation rate G in a semiconductor of thickness d with an absorption coefficient α and a reflectivity R under illumination with an incoming photon flux Φ_0?

8.10 What is the closest value for the "Shockley–Queisser" limit for the efficiency of an ideal solar cell under illumination with unconcentrated sun light?

(A) 12%
(B) 19%
(C) 30%
(D) 37%
(E) 42%

8.11 Sketch the equivalent electrical circuit of a real solar cell.

8.12 What is the formula for the open-circuit voltage U_{oc} of an ideal solar cell?

8.13 What is formula for the I-U characteristics of a real solar cell under illumination, including series and shunt resistances R_s and R_p?

Problems of Chap. 9

9.1 What is the figure of merit for thermoelectric materials?

9.2 According to the Boltzmann approximation, what is the Seebeck coefficient of an electron-like semiconductor?

9.3 Sketch and describe the current-voltage characteristics of a thermoelectric power converter.

9.4 What is the figure of merit ZT of a thermoelectric material with a Seebeck coefficient $\alpha = 0.1$ V/K, an electrical conductivity of $\sigma = 1$ A/(Vm) and a thermal conductivity $\kappa = 10$ W/(km) operated at 100 °C?

9.5 What is the Seebeck coefficient of an electron-like metal? Include at least all terms important for the correct unit of α_{met} and the correct sign.

Index

M. Stutzmann and C. Csoklich, *The Physics of Renewable Energy*,
Graduate Texts in Physics,
https://doi.org/10.1007/978-3-031-17724-8